ROOTS AND RESEARCH IN Urban School Gardens

This book is part of the Peter Lang Education list.
Every volume is peer reviewed and meets
the highest quality standards for content and production.

PETER LANG
New York • Washington, D.C./Baltimore • Bern
Frankfurt • Berlin • Brussels • Vienna • Oxford

VERONICA GAYLIE

ROOTS AND RESEARCH IN Urban School Gardens

PETER LANG
New York • Washington, D.C./Baltimore • Bern
Frankfurt • Berlin • Brussels • Vienna • Oxford

Library of Congress Cataloging-in-Publication Data

Gaylie, Veronica.
Roots and research in urban school gardens / Veronica Gaylie.
p. cm.
Includes bibliographical references.
1. School gardens. 2. Urban gardens. 3. Urban gardening.
4. Environmental education—Study and teaching. 5. Urban schools—Research.
6. Education, Urban—Research. I. Title.
SB55.G395 372.35—dc23 2011013895
ISBN 978-1-4331-1526-4 (hardcover)
ISBN 978-1-4331-1525-7 (paperback)

Bibliographic information published by **Die Deutsche Nationalbibliothek**.
Die Deutsche Nationalbibliothek lists this publication in the "Deutsche Nationalbibliografie"; detailed bibliographic data is available on the Internet at http://dnb.d-nb.de/.

Sketches of Eldorado Elementary School (above) and Grandview/Uuqinak'uuh School Garden (below) by Tracy L. Penner.

The paper in this book meets the guidelines for permanence and durability of the Committee on Production Guidelines for Book Longevity of the Council of Library Resources.

29 Broadway, 18th floor, New York, NY 10006
www.peterlang.com

Printed in the United States of America

For
poets in gardens,
everywhere.

Tell the story of your beets and how they grew.

—Clarence Weed

CONTENTS

ACKNOWLEDGMENTS

I am grateful for the many people who helped in the process of creating this book. I thank the generous garden educators and volunteers who took time to participate in various ways in this project, including: Maria Acosta, Cyrene St. Amant, Christine Boyle, Opal Essence, Lori Fetzer, Linda Gregory, Dave Goodward, Janeen Hobbs, Marcia Ingersolv, Brent Mansfield, Russell ("Mr. Russ") McCall, Gretchen Noble, Tracy Penner, Illene Pevec, Susan Schellehamer, Bonnie Trusler, and Ydalia Yado. Special thanks to Sharon Vieira, teacher at Grandview/Uuqinak'uuh, and her students. I also thank the following administrators for their support of school gardens: Adelina Aramburo, Patricia Hunter, Liz Kloepper, Karen Hoag, Gloria Raphael, and Tai-Sun Schoeman. Thank you to Mark Francis at UC Davis, and Rebecca Martusewicz at Eastern Michigan University, for the thoughtful suggestions and encouragement when most needed.

Thank-you to all of my colleagues at the University of British Columbia in Vancouver, and elsewhere, who provided support, collegiality, inspiration, feedback and friendship while I worked on this book.

Thank-you to my colleagues and friends at the Walking the Talk Network/ Institute for Environmental Learning/United Nations Regional Centre for Expertise (RCE) at Simon Fraser University.

Special thanks to Chris Myers at Peter Lang Publishing in New York for his ongoing support and dedication to innovative research. Thanks also to Bernadette Shade, for her expert production guidance.

With gratitude to my mother Grace, who teaches me about generosity and soul, everyday. To my sister Monica. To the boys, Louis and Rémi.

And thanks to Chris, for the constantly cool spirit, and the lovepower. *Mo Anam Cara.*

Other acknowledgments

Thank-you to the following individuals who granted permission to reproduce their work:

Photograph of *In the Moment* by Rick Carpenter (2002) at the San Francisco (SF) Recycling and Landfill Sculpture Garden was reprinted with permission from the artist. All rights reserved.

Grandview/ Uuqinak'uuh School Garden Design Plan and other original sketches by Tracy Penner, Landscape Architect, were reprinted with permission from the designer. All rights reserved.

Please note: all student writing is anonymous and complies with guidelines for The University of British Columbia Research Ethics Board for Research with human subjects.

· 1 ·

INTRODUCTION

I love it.
I love the garden.
That's what I want the world to know.
(Student, El Dorado Elementary School)

The urban school garden represents a cultural, pedagogical and practical shift in priorities towards the earth. As a small-scale response to contemporary environmental issues now facing the planet, the widespread rise in urban gardens is also a highly visible sustainable practice that not only includes local schools, but the communities, regions and histories that surround them. As a form of urban agriculture, the tangible cultivation of local place offers a source of food and a connection to the earth that inspires new ways to think, learn and live and invites new connections between school and community. In cities, such places are urgently needed. As cities now attempt to address the growing food crisis, school gardens provide an opportunity to address, in small and effective ways, "the crisis in how we think." (Orr 2006)

This book explores the value of an urban school garden as a bridge between practice, pedagogy and community. The cultural significance of an urban garden is explored from its historical origins in industrial cities through to its current manifestation in four regions in the Pacific Rim: Southern California;

Northern California; the Pacific Northwest; and Southern Coastal British Columbia. All of the gardens in this book are small. And all are linked to a larger history of urban school gardens that has, in one form or another, connected learners to the land.

The Need

The need for locally produced, urban food has dramatically increased in both developing and industrialized economies. Currently, three-quarters of global population growth occurs in urban areas. By 2020, the urban population in developing countries is expected to double to nearly 4 billion, with eight of the nine mega-cities increasing populations in excess of 20 million. (Desmond, Grieshop & Subramaniam 2004). As populations rise, as agricultural land disappears, as food security becomes an even more urgent and complex issue, environmental education has become an urgent area of research. In addition to the most visible purpose of a school garden (to grow food), UN research documents report that other positive benefits of school gardens (access to local food production and training; increased learner engagement; student connection to the land) are doubly needed in urban areas. If urban communities in any country are subject to the challenges of poverty, hunger and over crowding, then it becomes even more important to involve students directly in the use and governance of their environments, in part through earth-centred teaching, learning and pedagogy.

Support for small-scale sustainable food production modeled in school gardens has emerged from a wide range of community interests as diverse as the gardens themselves. While the learning benefits are apparent to many who work and learn in such places, the recent school garden movement gained widespread popularity through both a desire, and a practical need, to grow and learn about food. In North America, political, socio-economic and health issues surrounding industrialized food have been the basis of popular publications that include: *Fast Food Nation* by Eric Schlosser; *Food, Inc.* by Peter Pringle; and *Food Politics* by Marion Nestle. A rapidly increasing number of community movements have also sprung up in the last decade that collectively promote social and community organizing around urban food issues at the local, civic, state/provincial and national levels. The list of such organizations continues to grow, and includes: Community Food Security Coalition; Food Secure Canada; Roots of Change; Tilth; San Francisco Green Schoolyard Alliance; Center for Urban Education on Sustainable Agriculture. One group in particular, Slow Food, now with offices around the world, specifically seeks to reconnect people to food beyond goals of production and consumption. This and other groups that back the school garden movement seek environmental and social sustainability, changes in food policy, and a larger transformation of cultural approaches to food.

Sustainable food advocacy groups from a variety of sectors have grown to form a political movement with food as the basis. In 2009, at the White House in Washington, DC, First Lady Michelle Obama started a vegetable garden on the White House lawn to help teach school children about local food and growing techniques. The same year, First Lady of California Maria Shriver installed the first edible garden at the California state capital. Since Shriver moved to the capital in 2003 the number of California school gardens doubled to 6,000. While it may appear that food has suddenly risen on the popular and political agenda, in fact, in addition to the brief flourishing of ideas in the 1970s (including Frances Moore Lappé's book *Diet for a Small Planet*) the roots of the recent phenomenon in fact began closer to the turn of the last century. At that time, multiple levels of government that included Education, Agriculture and later, the military, banded together to support one another's interests around food. The first movement gained momentum with war and food crises in a way that mirrors today's sense of urgency around climate change, rising populations, food security and shrinking farm land.

While current interest runs high, somewhere between the flourishing of the early 20th century movement and current interest, school gardens lapsed for decades. During this time the movement was maintained in small pockets by informal urban community groups and by a variety of leaders from outside the field of education who persisted in their support for environmental issues. History shows that the survival of school gardens has always relied on a diversity of community support. A history of urban gardens also reveals that access to limited green space in urban industrialized cities specifically requires the work of both reform leaders and informal urban networks. Currently, the ideal of a "green economy" (Jones 2008), and the inclusion of previously marginalized groups in sustainability projects in industrial economies has also emerged as a core interest that promotes values of social justice, equal opportunity and access to urban land.

In schools, the actual integration of school gardens into lessons and curriculum naturally varies between regions, districts, schools, and individual classrooms. The inclusion of lessons around food is still typically linked to subjects around health and science. At the middle school level, teaching about food and gardens is linked to research that supports the cognitive, social and emotional benefits of experiential, interdisciplinary learning amongst pre-adolescent learners. (Grant & Littlejohn 2004) Wide popularity and support for school gardens still largely rests on the linkages between the gardens, curriculum, and government standards. A few key groups promote study and practice in ecology-based education that integrates principles around sustainability, food security and Slow Food across school curriculum. The Edible Schoolyard (founded by Alice Waters, also the Vice-President of Slow Food International) provides a national model school garden for integrating food and sustainable food principles into local school projects and curriculum. The Centre for Ecological Literacy is a also a leading foundation in this area as it focuses on the development of deeper understanding of the ways that the land may be linked to learning in school curriculum. Groups that support research and practice, from a wide range of teacher, researcher and community advocates, are essential to the development of pedagogy and to understanding the deeper changes in how humans think about, dwell within, and learn from the earth.

As research, school gardens can be linked to learning values around experience, freedom and empowerment. After the decline of the first reform movement in the early 20th century, gardens cultivated a strong, diverse community base where it remained alive in small pockets and from which the current trend now rises. As a movement, school gardens remain rooted in community, schools, local need and interest, while academic research in the field potentially bridges understanding between practice and pedagogy.

The Evidence

Evidence is overwhelmingly positive in the ways school gardens contribute to students' academic achievement. A recent study from the UK's National Foundation for Educational Research sponsored by the Royal Horticultural Society (2010) surveyed 1,300 teachers and studied ten schools to examine the impact of gardening on students. The results reported improvements in student social and problem-solving skills, including a dramatic improvement in literacy and numeric rates. The report examined how school gardens stimulate a love of learning through the practices of experiential learning, which inspires curiosity, movement, and helps students translate academic subjects into real life practice. The report also included evidence that students in gardens learn from the changing natural elements in a way that encourages flexibility, the ability to face fears, and adaptation to thinking and learning in diverse settings.

Other foundational studies indicate that the benefits of school gardens, as a progressive, cultural teaching tool, would also uniquely benefit gardens in urban contexts. For example, the Learning through Landscapes Project, a major UK study, found that the impact of school grounds, both positive and negative, have an impact on student learning and attitudes about their total school experience: "school grounds...convey messages and meanings to children that influence their attitude and behaviour in a variety of ways...Children read these messages and meanings from a range of signifiers, which frame the cultural context of the environment" (cited in Evergreen Foundation Report 2000b, 2). Almost one-third of Britain's thirty thousand schoolyards are involved in the program, which inspired a similar program in Canada called Learning Grounds, and a major Swedish program, Solons Uterum. (Louv 2008)

Canadian studies on school greening projects have all supported the positive health benefits of creating gardens in school grounds. One large-scale study (Dyment 2005) examined information provided by 105 parents, teachers and administrators at 59 elementary schools in British Columbia, Alberta, Manitoba, Ontario, Quebec and Nova Scotia. Results indicate that green school grounds play a significant role in promoting physical activity for students of all ages, interests and abilities. Another Canadian study (Dyment 2006) noted, among other findings, that in using school grounds:

- Teachers were able to deliver a broad range of subjects in the outdoor classroom created through greening projects;
- Student learning was enhanced;
- Teachers had renewed enthusiasm for teaching and were using a wide variety of innovative instructional strategies on green school grounds;
- Students demonstrated more positive social behaviour when learning and playing on green school grounds;
- The diversity of play spaces created through greening projects suited a wider array of students than conventional turf and asphalt school grounds;
- Green school grounds promoted the social inclusion of all people, irrespective of gender, race, class or intellectual ability.

In the United States, a nationwide survey examined attitudes about Environment as an Integrated Context for learning (EIC) (Evergreen Foundation Report 2000). The study included the experience of students, teachers and administrators in 40 elementary, middle and secondary schools that include the natural environment as a "curriculum element." (4) Further to surveys and interviews on school programmes, standardized test scores for students involved in EIC showed consistent improvement in reading, math, science and social studies. In another study, over 4,000 school principals were interviewed on the use of school gardens in academic instruction. The study noted a strong need for curriculum materials, teacher training, funding and nutritional knowledge in creating successful gardens (Graham, Lane Beall, Lussier, Mclaughlin & Zidenberg-Cherr 2005). Other researchers report similar findings, noting that the success of experiential, garden-based, project-based learning relies on further research and on the development of ecology-centred curriculum frameworks (Desmond, Grieshop & Subramaniam 2004).

A study commissioned by the Centre for Ecoliteracy at The Edible Schoolyard (Murphy 2003) examined the garden's holistic pattern of education for sustainability, which includes ecological knowledge, cooperative school culture, immersion in a hands-on project and learning based on a sense of place. The study also investigated students' knowledge and attitudes towards health, nutrition and agricultural knowledge. Findings included a variety of positive outcomes, including the effects of school gardens on academic achievement and positive impact of the notion of education for sustainability on student psychosocial understandings. Some conclusions of the study indicate improvements in "understanding garden cycles," "psychosocial adjustment," "student sense of place," "understanding of sustainable agriculture" and "increased knowledge of the food students eat" (6). The study also noted that the teachers who taught in the schoolyard "ranked teaching academics, cooperation, and compassion for living things as their three highest priorities" (Murphy 2003). Research commissioned by the CEL (Stone 2005) reported improved student behaviour, health and increased knowledge of ecological literacy.

Several research studies further support the value of experiential, hands-on, eco-centred learning as a means of improvising performance on standardized achievement tests (Klemmer et al. 2005; Smith and Motsenbocke 2005; Lieberman & Hoody 1998). While more research is needed in the area of the specific impact of school gardens on urban student learning at all levels, studies (Waters 2008; Murphy 2003) show that learning through gardens improves overall academic achievement and student engagement—two areas in which urban learners might uniquely benefit. UN researchers (Desmond, Grieshop & Subramaniam 2004) strongly support a global need for more eco-centred curriculum, for further teacher training in garden-based methods, for more community involvement around environmental issues and for further study on the academic and social effects of GBL.

Improved test scores only begin to describe the value of eco-centred, experiential learning at all levels of curriculum. For school gardens, particularly urban school gardens, to gain widespread acceptance as a site for critical, educational change, educators (and their institutions) must also approach their gardens beyond the context of state standards. A school garden is potentially the start of a groundswell of movement for teachers ready to engage students in experiential, transformational environmental learning. Learning in a garden seems to somehow change evaluation criteria as students move beyond teacher, and even research, expectations. The real value of learning in gardens is more difficult to measure and requires further research, and perhaps even new ways of measuring the value of such research.

UN reports suggest that, globally, the benefits of school gardens as a form of ecology-centred teaching are diverse. In reviewing their extensive findings researchers conclude that a garden pedagogy should include a wide range of skills development, which includes academic skills ("enriching core curriculum"); personal development ("a sense of excitement, adventure, emotional impact and aesthetic appreciation to learning"); and social and moral development ("To teach sustainable development; To teach ecological literacy and/or environmental education; To teach the joy and dignity of work"). (Desmond, Grieshop & Subramaniam, 49)

As an example of the evolution and new, emerging diversity of garden-based curricula the Toronto District School Board uses three broad concepts in integrating environment to established curriculum: Sense of Place; Ecosystems Thinking; and Human Impact. Such holistic measurement includes open-ended ideas around environmental learning such as learning how human systems are interconnected and interdependent. Using a critical lens the new curriculum also advocates "weighing the consequences, both helpful and harmful, of human interventions into natural systems." (Cited in Houghton 2005, 20–21) The British Columbia Ministry of Education Curriculum Supplement *Environmental Learning and Experience, an Interdisciplinary Guide for Teachers* (2007) similarly suggests designing curriculum around holistic principles that include: integration of subjects; critical reflection on a range of perspectives; a focus on issues from local and global perspectives. The same report recommends that environmental learning occur within contexts of "hope" and "humility."

The recent curriculum design guides emphasize an organic approach where learning evolves within local and communities. New curriculum is strongly influenced by global environmental research from the UN that emphasizes how the effective integration of ecology in schooling begins with high ideals, broad visions and hopeful perspectives. New environmental learning describes pedagogical goals such as creating hands-on connections between students and their natural surroundings, while including new media literacy and traditional knowledge. Recent case studies of urban community gardens reveals similar lessons for urban school gardens research. (Hou, Johnson & Lawson 2009) Such studies promote the idea of community gardens as multiple expressions of urban sustainability that always include the diverse perspectives of planners, policy makers, designers, volunteers, community and professional involvement.

Historically, in mainstream schooling, teaching *about* the land usually did not include learning *within* the land (and vice versa). Up until quite recently, research has taken a similarly hands-off approach. The teacher's (and researcher's) role in an urban school garden, in addition to facilitating ecoliterate starting points for curriculum, now involves working alongside students and teachers in community. Particularly in urban settings, the establishment of sustainable models for learning now involves land stewardship where teaching *about* the environment includes firsthand experience within that environment. New garden research studies and curriculum models show a future where students will learn to understand and nurture ecological systems within the actual experience of those systems.

The Larger Benefits: A Culture of Learning in/as Place

A pedagogy of school gardens benefits from research that combines practice, environmental education philosophy and describes how a garden combines both in a way that profoundly stimulates learner imagination. While research clearly supports the learning value of gardens, few studies integrate the deeper values of environmental education with sustainable learning practices. The notion that place itself always forms part of understanding is a concept not typically acknowledged in Western models of learning. Apart from a few notable examples from environmental design (Francis & Hester 1990) discussions around the value of place tend to focus on practice and on the study of gardens as a curricular or learning enhancement tool.

While urban gardens today usually begin with practical needs around food, the cultural and learning value of such spaces are even more far reaching. Urban school gardens provide a practical focal point for shifting ideas around environment, pedagogy, land, planning and urban issues. Resurgent interest in food gardens is taking place in urban schools (and parks, empty lots, public land

and unused private land) because such places represent a portion of the dwindling public space left in inner cities. School gardens in urban areas can be as small as one garden box and a few seeds; other projects involve entire schools, requiring extensive planning and investment. No matter the size of a school garden such spaces potentially revolutionize learning in part by changing previous perspectives on land.

Urban farming, like urban schooling, presents challenges unique to its location; the odds of growing food in such a limited space alone redefine the concept. The current necessity to cultivate urban gardens and farms also presents an opportunity to teach students lessons of farming that translate to their needs as urban learners: shared work; community and self-sufficiency; practical training and application of knowledge; working through adverse circumstances; empowerment through stewardship; engagement with earth's systems; connection to the land. When applied to the city, lessons of farming become linked to the values of urban learning and to the particular survival needs of a community. In the city, school farming involves inspiration, a cultivation of care, as well as the empowerment that comes with taking action and implementing technical know-how.

As place, and as pedagogy, a garden embodies an "earth democracy" (Shiva 2006) involving an ethic of justice necessary to the earth's survival. A school garden supports the well-being of citizens and promotes biodiversity and local action and brings a workable solution to globalized monoculture and socio-economic injustice. The phrase "all education is environmental education" (Orr 2004, 12) emphasizes learning in (and *as*) the processes of the earth; such a concept is perhaps even more important in the inner city where ability to negotiate environment brings survival. Environmental educators (Bowers 2006; Shiva 2006) describe how industrial models of education, over many decades, promoted unsustainable learning that focused on a concept of individual human success. Such models supported strict, generic, narrowly defined learning outcomes that had little to do with environmental contexts. In the city, the idea of being severed from nature as an inevitable function of industry now requires a renewal of commons ideals around shared space and mutual support. Industrial culture, based on rationalist traditions of knowledge, severs human connections to the land and impedes the potential towards self-sufficiency of the commons. Industrial traditions that deny living, shared support mechanisms of thought, practice and place threaten democratic process and long-standing traditions of survival based on a commons conservation model of mutual support and self reliance. (Bowers 2006)

In North America a perspective on land as capital has also had a long historical presence. The social and cultural values of gardens in cities are also linked to a deeply ingrained Renaissance worldview where nature was perceived as being the lowest rung of a strictly defined, linear hierarchy. (Benton-Short & Rennie 2008, 24) Such perspectives permeated western society, design, and learning at all levels. The geometry of gardens (typically squares separated by straight pathways) was a further symbol of formalized barriers within the organic flow of nature. In Europe, the scientific revolution in the 16th and 17th centuries created a perspective where nature was viewed as an extension of human ingenuity and the totality of human mind and nature was seen as a machine necessary for both capitalist production and as a means of solidifying the social order. At the turn of the last century industrial design ideologies persisted in school gardens with a focus on individualized plots and strict institutional control over learning outcomes. The physical shapes of gardens again symbolized learning that was individualized, exclusive, uniform, and easily controlled. Over a century later, cultural programming from industrial eras not only influences what is learned but how and where it is learned.

Changes in the environment now call for changes to the ways nature is perceived in Western learning. New perspectives that honour the environment are inseparable from goals of sustainability and will increasingly influence and redefine society, design and schooling at all levels. As the population of urban poor continues to increase, school gardens provide a place for vocational farming and for promoting ideals around stewardship, equality and access. The potential for the transformation of industrial urban space expands as vacant lots give rise to community gardens. In Philadelphia, for example, 500 community gardens have been planted on land that now involves over 3000 families. Ultimately, vacant land claimed and shared by local community as food gardens represents the reversal of the industrial enclosure process. (Bowers 2006, 101). The return to tradition also represents a practical revolution that extends the work and cultural significance of early reformers, who sought to revitalize the commons over one hundred years ago.

Urban gardens are inseparable from broader socio-cultural contexts. Current movements (Van Jones 2008) now address the problems of the planet alongside the problems of the poor, especially in the inner city, with the integration of lower socioeconomic groups into the economy of sustainability. Leaders of the "Green Economy" movement suggest that a solution to the problems of environment are found amongst the unemployed:

> Let us all say together: 'We want to build a green economy strong enough to lift people out of poverty. We want to create green pathways out of poverty and into great careers....We want this 'green wave' to lift all boats.' (16)

Projects that bring ecological and social justice to urban public space are on the rise, now most notably in Detroit, where the revitalization of unwanted land in the inner city lead to a foundation for several innovative community-based projects. There are, in fact, now over 1000 community gardens in the city, many of which are linked to schools or to informal learning. In one project, at the Old Redford neighborhood in Detroit (Martusewicz 2006), local facets of the commons (land, language, culture, tradition) are all nurtured and studied in urban gardens. In another project, Grace Lee Boggs' (2009) "Detroit Summer" began in 1992 as an intergenerational youth movement to develop community gardens in the post-industrial city. The project included African American elders (called "gardening angels") who worked with local youth on projects that eventually grew into a large urban agricultural movement with neighborhood gardens, youth gardens, church gardens, school gardens, hospital gardens, senior independence gardens, wellness gardens and Kwanzaa gardens. Part of the project also includes the work of Capuchin monks at Earthworks, a program centered around gardening and Detroit school children to help youth learn about organic farming, science and nutrition. The garden provides fresh produce for the Capuchin Soup Kitchen and for the WIC (Women, Infants and Children) supplemental nutrition program. At the Catherine Ferguson Academy, a public high school for pregnant and parenting girls, an urban farm was created to help provide fresh food to impoverished teens. Another organization is the Detroit Black Community Food Security Network, which developed a number of programs around urban agriculture, public policy, and vocational training in environmental-based careers. The group started D-Town Farm, a two-acre model urban farm and The Food Warriors Youth Development Program, a project that links elementary school children to agriculture. The schools work with the Southeast Michigan Stewardship Coalition to provide community based, contextual research that includes those working at the forefront of food justice in the inner city. Democratic, local participation in learning thus becomes part of practice and pedagogy where ecology is interpreted as an inclusive, complex network of living relationships embedded in social justice and community well-being. The garden projects around the city of Detroit have been described as "...a testament to the fact that even the poorest of the urban poor can envision an alternative future—and can mobilize the community." (Boggs 2009, 48)

Small urban plots require innovative use of scarce resources that provide a global model for sustainable agriculture. Given the urgent need facing the urban poor, such communities are now leaders in developing sustainable resource methods. Such work also represents an ideal rooted in a long historical network of grassroots urban reform that inspires and provides a basis for others to move forward, or even begin, in small ways.

Multiple Methods: Field and Philosophy

> What does nature mean to us? Yosemite: Majesty. A sacred place. The desert: Peace. Harshness. Clarity. Songbirds. Beauty. Delicacy. Earthquakes: Fury and vengeance. Water: A metaphor for life. But nature? The overall meaning that frames all others…is wildness. The not-modern. The not-us-ness. What's out there. Refuge. Salvation. These meanings historically have reigned…(Price 2005, 227)

This study examines the holistic, practical and philosophical meaning of an urban school garden. The gardens chosen for this study are small. As educators try to address problems now facing the planet this study also attempts to make research more accessible to local communities and to teachers with little space, and typically dwindling resources. In such a climate, firsthand stories in small spaces are necessary to understanding the communities we live and learn in and to understanding how responsible use of small space might potentially solve global needs.

As urban farm, as community, as pedagogy, urban school gardens serve multiple purposes and thus, carry multiple meanings. And, while the sites themselves are small, and the work practical, the collective benefits are large. There is also a long-expressed need for new ways of exploring environmental learning through gardens as place, as meaning, "…and as an ecology of interrelated and connected thoughts, spaces, activities and symbols." (Francis & Hester, 1990, 2). Thus, new, holistic interpretations of garden learning bring the need for multi-faceted, multi-sectoral approaches to the complex problems facing the earth; such a purpose requires a blend of research, practice, philosophy, pedagogy and recognition of historical practices. And, while many early urban gardens existed solely for aesthetic purposes, today's gardens invite a multitude of new methods to teaching, learning and research.

A garden is a way of both making order and of making what one teacher called "a beautiful mess." Since the garden is always "a complex ecology of spatial reality, cognitive processes and real work" (Francis & Hester 1990, 8) holistic approaches to garden research are now necessary in order to truly understand the value and complexity of such spaces. Thus, this study explores urban school gardens through the gardens themselves and through the history, philosophies, cultures and practices of the surrounding community. The study examines the realm of school gardens in urban communities through methods that include gathering the first-hand stories of students and teachers: the experts with dirt under their fingernails. Here, expert pedagogical and design theory combines with the story of how teachers and learners actually interact with their local urban environments. The study seeks to engage in new pedagogy, in new ways of creating balance between practice and philosophy that includes voices from the field.

The methods used in this project are meant to contribute to a holistic understanding of school gardens. In each chapter, the history, philosophy, practice, story and analysis of urban school gardens are first examined as a whole and then, specific to each case study/region. For this study, I specifically sought out school gardens where the garden: grew in a relatively small space in a dense urban area; was in a stage of growth and development; was strongly supported by administration; and where the garden was a central feature of the school. In short, I sought out small plots with high hopes.

In approaching these gardens, I also wanted to ask new questions that would help bring a new, perhaps more subtle understanding of such places. At each garden, questions were asked about: the garden history; the unique benefits of the garden for their specific urban area; how the garden influenced teaching and learning. In the interviews I also invited the teachers to speak freely about their ideas, perceptions, notions, visions and dreams around the spaces. For the teacher interviews it was often not until later, during post-interview visits and follow-up emails, that the teachers began to define the meaning of their particular garden.

The discussion of the case studies is framed in historical practices and a chronology of the evolution of garden pedagogy. Each chapter describes the gardens through a brief history of the interaction between community groups and land specific to the urban region, followed by: interviews with educators and students; samples of student writing; and observations on story and theme. At each site methods included: on-site interviews and observations over a series of lessons, and questionnaires with entire class groups. At each of the schools the teachers and I then asked students to describe their experience of the garden, in their own words. Students responded to a series of questions about their attitudes and beliefs about their gardens and were then asked a series of open-ended questions. (What do you like best about your garden? What would you like to tell the world about your garden?) The case studies were examined for the different ways that students found connection to the earth: how do students learn local gardening practices that help sustain the land? How do educators and students describe those connections? The study also explored the underlying philosophies of the garden inspired by educators and planners that in turn influenced the pedagogical possibilities of each space. At every garden I asked educators *what* was learned in the garden, *how* it was learned and whether they thought the garden somehow created a magical space for learning.

In reviewing the texts, the student responses were often passionate, poignant, and typically included a disarming honesty to questions about both their garden and the natural environment. In speaking about nature, the student language, at all grade levels, took on a lyric, poetic quality that often included repetition and positive emotion. As the students expressed themselves about nature, some of their responses almost appear as short, spontaneous poems.

Every story in every garden is unique. The project allowed for a diversity of environmental and cultural themes to emerge from each space. At one site (the Grandview/Uuqinak'uuh garden in Vancouver) the project evolved from student poetry in the school garden to cultural expression of nature to the needs of local urban environments. Student responses at all of the schools, whether expressed through interviews, poetry or illustration, bring to light a record of student connection with their own habitats. The students all spoke of pride, stewardship, community, facing fears, and connecting their knowledge to their local environments and beyond.

The research attempts to describe the ways that learning and connection is found within gardens and in the places, communities and cities where they grow. The study focuses on the benefits of school gardens as told through the historical and current stories of educators and students who envisioned, work within, and learn from those spaces. Various threads are explored (historical evidence, current philosophy and practice) in a holistic study of the ways that urban school gardens contribute to the larger, overlapping stories of philosophy, pedagogies and practices of environmental education. This study thus seeks to explore the complexity and overlap between philosophy and practice, between ideas and physical labour, between urban life, the natural environment and the stories of small spaces. The study ultimately describes how ground-level land stewardship in urban schools leads to deep transformation in pedagogical practice.

The multiple meanings of gardens (as symbol, knowledge, practice or pedagogy) provide new perspectives on learning. Student/teacher stories, in the context of a much larger narrative, illustrate the value of the gardens themselves and show how the inclusion of smaller stories and meta-narrative changes the criteria for interpreting value. Overall, the inclusion of diverse methods and multiple forms of analysis not only illustrates the philosophical and practical dimensions of such sites, but also illustrates how gardens are both *used* and *thought about* by those involved at the ground level.

Where Learning Begins

An urban school garden represents a practical and powerful symbol of nature in the city. The way land is used in cities (as park, as commons, as private land, as shared garden) shapes community, culture, and learning, over centuries. From turn-of-the-20th-century reformers to current interest in gardens, the notion of developing urban areas for farming and education touches on larger ideals around community, equality, and access. While suburbs grow around cities as sites intended to promote both rural and urban ideals, the city itself is increasingly viewed as its own ecosystem. An urban school garden symbolizes both social and environmental awareness; how a region values and preserves its common green space defines the underlying values that guide a community's approach to land stewardship. While there is much still to learn, an urban school garden is emerging as a symbol of place between urban and rural ideals that provides a bridge between both.

School gardens are rapidly expanding in North America. And while there is growing amount of research evidence and practical information, less has been recorded on the pedagogical, historical, design and cultural aspects of such sites. The ways that gardens potentially transform learners and the learning process has become part of the conversation. There is urgency to develop *urban ecological literacy* that starts with students' firsthand awareness of their local environments. There is need to move toward small, effective solutions that will enhance urban systems and potentially transform communities. A new model of ecology-based learning focuses on the diverse needs of local place and cultivates attentiveness, observation, involvement, equality and respect as a basis for sustainability.

Sustainable cities begin with a recognition and valuing of place. A garden in an urban school is *a garden*, a place to grow vegetation, and also a cultivation of the connections that link students to their own survival. As a bridge between cities and nature, between philosophy and practice, gardens bring value to the complex knowledge and practices of the local environment within schooling. In researching this book, what began as a story about school gardens became a story about the cultural and educational value of place, from industrial cities in Eastern states, to urban deserts in Southern California, to the inner city of Vancouver, Canada. In the urban garden, learning begins in small scale. It begins where students begin. At their front doors. In their school yards. In their imaginations. Ultimately, student connection with the land brings about the social, ecological and cultural knowledge necessary for sustainable learning.

· 2 ·

URBAN SCHOOL GARDENS

History, Pedagogy, Practice

The question of space is a serious one.
(Klein 1904)

The history of urban school gardens is influenced by late 19th-century urban reform, post-war Victory Gardens, late 20th-century environmental change and early 21st-century grass roots movements. From early factory gardens in New York City to The Environmental Schoolyard in Berkeley the evolution of urban gardens reveals the pedagogy, practice, and social reforms that surround such places and that continue to influence the current movement.

A pedagogical history involves both chronological events and thematic accounts and includes firsthand perspectives of educators from the popular journals of the day. Various pedagogical threads, from a variety of perspectives, reveal how past and current ideas potentially contribute to current innovations. In the historical evidence, as well as the case studies, the smaller stories become part of a larger historical theme that includes the ways that students have encountered nature in the city as central to their learning. An investigated history of urban gardens includes varied approaches which help uncover the deeper cultural meaning of such sites where

> The power of the garden lies in its simultaneous existence as an idea, a place and an action. While each has a value as a way of thinking about gardens, viewing them together offers a deeper, more holistic perspective on garden meaning. (Francis & Hester 1999, 8)

In the early 20th century, nature and teaching journals were authored and read by those at the core of the urban reform movement. Such journals and other print materials now emerge as valuable research documents that help illustrate the research, practice and pedagogy of the time. First person narratives, which may or may not present a factual account of events, nonetheless presents the lesser-known stories from those who actually worked in the field. The journals would have been read by researchers, practitioners, teachers, and decision makers and therefore reflect the views that informed popular opinion and ultimately laid the ground work for the pedagogy of urban school gardens. Their viewpoints represent broad perspectives of front-line practitioners in a way that begins to explain the roots of garden pedagogy.

The historical documents also include writing from people who would have been well known by educators across the country. C.D. Jarvis was the first director of the Bureau of Education's Office of Home and School gardening. M. Louise Greene's (1910) influential book, *Among School Gardens*, was an early, comprehensive account of teachers and learners using school gardens in the context of curriculum. Louise Klein Miller, author of the text *Children's Gardens for School and Home: A Manual of Cooperative Gardening* (1904), was the curator of school gardens for Cleveland Public Schools. Benjamin Marshall (B.M.) Davis, author of *School Gardens for California Schools: A Manual for Teachers* (1905) and *Nature Study in the Los Angeles State Normal* (1901) was also an influential educator in the early 20th century. Such accounts all helped develop the first initial ideas around school garden practice and pedagogy. While some of the ideologies change, or perhaps just sound new, the larger narrative of urban school gardens describes practical human initiative that helped shape cultural perspectives on how gardens are used, and thought about, today.

Historical Overview

Where, and when, did the first shared urban gardens exist? In the ancient world, about 90% of urban dwellers were peasants who worked in the local fields for food. The notion of commonly held, urban, community gardens would have been forbidden under strict Roman laws. Roman rulers developed and controlled every aspect of life in both urban and rural areas, including agriculture. Archeological digs have revealed that farming inside the city wall existed 4,000 years ago on the Indus, Tigris and Euphrates, and later, on the Jordan River. (Smit 2006) In medieval Europe, kitchen gardens flourished inside urban fortresses. During this time, urban gardens, and kitchen gardens (the gardens inside the walls) were increasingly recognized as a pleasurable place where community gathered. In Pompeii each household had its own gardens used not only for food but also as a central place for family to socialize. (cited in Williamson 2002) In Arab countries, gardens of fruit trees and ornamental irrigation, in pools and streams, were planned throughout towns and cities.

Gardens used for instructional purposes were put in place by the Persian King Cyrus, who, according to early 20th century documents, developed gardens to teach the sons of noblemen about horticulture. However, a lack of documented evidence about the pedagogical processes of urban gardens (in the Western world) persisted for centuries. During the early 16th century Italian universities emphasized the importance of urban gardens for integrating pedagogical and agricultural purposes.

The idea of learning in urban gardens was interpreted in various forms over the centuries. In France, Rousseau (1712–1778) described the importance of school gardens as a model for learning and not simply as a tool for instruction. Francke (1663–1727), who established an orphan asylum in Halle, developed a school garden, where children were encouraged to spend leisure time. (Davis 1905) Some educators described the pedagogical possibilities of gardens early on: "School gardens have been laid out, neither to draw the attention of passers-by, nor to give great returns, but to instruct." (cited in Klein 1904, 11) From a social reform perspective, Pestalozzi (1747–1827), a farmer and educator, emphasized the purpose of the garden as an agricultural and education site for students who lived in urban orphanages. He established the farm as a central point of combined agricultural and educational training, where outdoor learning could help create a balance between "hands, heart, and head." (Desmond, Grieshop & Subramaniam 2004, p. 2) Froebel (1782–1852) gave employment to older children in his kindergarten and led the movement for school gardens in major cities across Germany. (Some German states encouraged school gardens as long ago as 1818.) By 1904, school gardens were policy in most European countries.

School gardens flourished in the early 20th century in North America and were seen as a means of promoting the ideals of progressive education and traditional notions of stewardship. The very first gardens in North America were started with a goal of pedagogical innovation; the first urban reformers saw the rise in school and community gardens in Europe and wanted to imitate the same projects back in America. In 1890 Henry Lincoln Clapp returned from Europe to install a school garden at George Putnam School in Roxbury, Massachusetts.

Putnam School was the first organized school garden movement that involved a coalition of school administrators, teachers, students, and volunteers; the garden lasted 30 years. Urban school gardens in the early 20th century were thus part of an expansive, diverse movement with shared involvement from educational, governmental, agricultural and community organizations. In 1910, a report to an agricultural education conference estimated an approximate 80,000 school gardens in the United States alone. (Welby 2003) Gardens at the time emphasized an aesthetic or higher moral purpose that would improve the perceived inferior living standard of urban life. In the late 19th century, the Director of the School Garden Association of New York wrote: "School gardens should be maintained by the city, the city owes it to the children whom it has deprived of breathing places and beauty spots through want of foresight." (cited in Desmond, Grieshop & Subramaniam 2004, 35)

At that time, while some educational outcomes were narrowly limited, other practices were clearly ahead of their time. For example, early on, many normal (teacher training) schools and agricultural colleges integrated their curriculum and practices around teaching and farming. In the early twentieth century, John Dewey referred to the promotion of agriculture in education as a "movement towards greater freedom and an identification of the child's school life with his environment and outlook." (cited in Ibid., 35) Dewey similarly supported learning within nature as a means of both understanding and nurturing organic systems. While Dewey did not so much describe the cultural or learning value of gardening *as work*, and while his approach to knowledge and schooling was very much based on a Western student success model, Dewey's practical theories made important links between historical daily life, the land and learning. He said:

> Gardening need not be taught either for the sake of preparing future gardeners or as an agreeable way of passing time. It affords an avenue of approach to the knowledge of the place farming and horticulture have had in the history of the human race. (cited in Subramaniam 2002, p. 2)

Ideas in the garden movement that brought real change and shaped pedagogy supported the principle that gardens were more than a teaching tool that served agricultural methods or political need; gardens in fact formed a base for social and academic reform by combining principles around environment, education and community. Such ideas, from Dewey and others, formed a precedent and practical basis for current approaches to ecology-based pedagogy in teacher education. (Gaylie 2009)

In 1918, practical interest in gardens rose as every state in America and province in Canada promoted school gardens as part of growing food for the war effort. After WWII Victory Gardens thrived; in subsequent decades school gardens declined as land previously used for gardens was taken over by athletic fields. (Subramaniam 2001) For a time in the 1960s and 70s school gardens rose in popularity again in part due to liberal social movements. Pedagogical support for gardens emerged from developmental psychologists who encouraged experiential education, hands-on learning and creative expression in a child's relationship with nature. (Cobb 1977)

In the 1980s urban school gardens faded in North America due to increasing socioeconomic conservatism and a growing emphasis on localized performance standards. In 1995, school gardens once again made a resurgence with former California state superintendent of schools Delaine Eastin statewide Garden in Every School policy that would stimulate school gardens across California. Behind Eastin's policy was a strong belief in the transformational aspect of school gardens, which she had witnessed in schools firsthand. She also believed in the power of gardens to break down the barriers of school-based practices to include local farms, thus connecting classroom learning to local (and global) context. (California Department of Education Report, 2002) Between 1995 and 2002 school gardens in California increased by 3000 and their numbers continue to grow. The current rise in interest in urban agriculture and education is at once practical, in sustaining the planet, even as such ideas around eco-centred learning slowly expand and transform mainstream learning.

The Roots of Urban Garden Pedagogy

Historical evidence reveals certain facts: the first organized school garden in North America was in Boston (1890); gardens at the turn of the 20th century in North America involved the cultivation of moral ideology and social control; from WWI onwards, school gardens were considered a practical necessity and source of food. Historical records also reveal complex cultural themes around urban ecology and about the way humans interact with land under changing, social, political and economic conditions. The history, however, is more than isolated stories about food, practice and politics. As a focal point for change the urban school garden is a unique laboratory of urban culture, ecology and society.

Urban school gardens, small cultivated lots in crowded cities, provide a microcosmic history of interactions between humans and land, of resourcefulness and persistence in the face of adversity, of cultivating spaces that might not at first seem to naturally lend themselves to agricultural, social, economic, ecological or cultural possibility. A history of urban school gardens is also a history of hard work on hard ground—both the work of physical labor and the work of imagination, of carrying out a shared vision.

In cities, the garden has been conceptualized, ritualized, and recognized for its unique, shared social value. Even a small garden in a city becomes a site for growing and learning amongst hundreds or thousands of people over time. The history of urban gardens reveals how humans have symbolized nature and how sites have historically been viewed as art, science, and symbols of learning that permeate all aspects of culture. The veneration of trees, for example, is central to many cultures and emerged from ancient Persian and Assyria, where a tree with a stream at its roots symbolized eternal life. (Spirn 1984, 184) Since trees were considered sacred, tree planting was considered a sacred occupation. Gardens were also viewed as symbols of plenty, a concept that politicians were able to use to prove their value among their subjects. In Nineveh, Mesopotamia, one ruler stated:

> I made gardens in the upper and in the lower town...with the earth's produce from the mountains and the countries round about, all the spices from the land of the Hittites, mer, vines from the hills, fruits from every country; spices and sirdu-trees have I planted for my subjects. (Ibid.)

A review of the evidence, over centuries, reveals a trend in the way garden pedagogy constantly shifts between lofty ideas and actual practice. The garden has long been a site of convergence for various groups with diverse, yet similar, agendas. Since gardening was viewed as a way to address a broad range of issues (educational, moral, social, and environmental) school gardens similarly received broad support from institutions, community, schools and government agencies. (Lawson 2005, 52) During the late 19th and early 20th century themes and practices in the urban school garden have included: training urban youth in agricultural knowledge in order to re-populate rural areas; a means of reinforcing mainstream moral or religious values; reclamation of the commons; and symbol of civic social responsibility.

Opposition to school gardens has taken place since the gardens started; such conflict has also helped shaped the culture of such sites. The simple notion of improving urban school life by improving access to nature immediately gave rise to competing interests. The argument over land turned to justifiable use and the most popular economic arguments usually won the day. Land was considered either developed, or empty; used or unused; profitable or unprofitable. Such ingrained ideologies have posed particular challenges to school gardens in urban areas throughout history.

In North America the history of urban school gardens also reflects deeply held perspectives on the historical separation of cities and nature. The story is one of escape and retreat: escape from the industrial city to the pastoral countryside. The city was not seen as a site that could cultivate learning in gardens since the city itself was viewed as radically separate from the perceived ideal experience of nature in rural life. In such a perspective, urban dwellers needed to be saved from a "degraded," inorganic, nature-less environment. As urbanites' relationships with the land developed into more practical pursuits such as small-scale farming in community gardens, gardens thus symbolized an important conceptual middle ground between the urban nightmare and the rural ideal. The symbol of a shared urban commons competed with another idea: when separation was not possible via rural land ownership or spacious urban mansions, more and more people could own a "pretend manor house" in the suburbs. (Kunstler 1993, 42)

Such perspectives were established in 17th-century Puritan America with the association between ownership and self-sufficiency. At this time, a colonial perspective on land as capital took root as American land laws were based on an overriding principle that land was a commodity to be used for capital gain. (Ibid., 26) Urban planning principles were based on a similar notion, where larger tracts of land were subdivided into smaller parcels of ownership, a method that continues to this day. Ever since land has been bought and sold for profit in capitalist societies land has symbolized ownership, domination and maximized use for profit. The ideas of land as the physical containment of community or social values, or commitments to the Earth in a transaction of survival, have historically withered under a capitalist ideology. Land and nature as ownership (ignored as shared resource, prized for monetary value) established a perspective where ethical stewardship was feared as a liability that threatened the right of ownership. All land was viewed through a lens of economic gain, and on whether variations in use, including shared ownership, somehow deprived an individual of the economic value of land. (Ibid.)

In the late 1800s some urban school gardens expressed controlled, miniature, urban versions of paradise that were meant to stimulate awe. These gardens were filled with observable symbols of Victorian beauty in the form of rocks, waterfalls, ferns and flowers. (*The Manufacturer and Builder 1894*, 205) By the early 20th century, an observational, nature-appreciation approach to urban gardens was viewed as antiquated. In a reactionary fashion, urban school gardens created primarily for observation were suddenly removed as wasteful sites of inactivity where children "….tired of walking around admiring the ferns, 'like old men and women' could "romp about and play." (Ibid.) The idea of simply learning by *being* in nature was not seen as useful in an era that emphasized ownership and a utilitarian approach to land. In short, observation and sharing were not viewed as legitimate forms of measurable experience that contributed to society. To be useful, land needed to be *used*. Nature itself was only to be claimed and used and was thus seen as separate from students' development as human beings. At the time, social controls on daily life severely segregated notions of work, learning, and the health/aesthetic benefits of nature. The notion that nature or traditional farming could actually help students learn in/as part of their environments and, further, that such interactions held potential links to agricultural vocations, was revolutionary for its time.

Perspectives on land extended to education, where teachers taught individual subjects in highly controlled classroom settings with students arranged, like plants in their garden, in rows. Purposeful activity did not include open-ended observation. Even the notion of recognizing urban *nature* was radical; any attempt to bring learners to the land, especially in the case of impoverished inner cities, thus required deliberate social intervention. Reformers in the early century wrote a great deal about promoting their ideas on both behavioral control and civic responsibility and many documents are filled with entreaties to aid the urban children in their cities. Since society itself was mired in values of social control, the documents spoke to people in a way that would spark a sense of social, civic, or economic duty. Reformers thus had to sway their audience using use the tools of the time (formalized language, print medium) to further their new goals.

Cultural Values of Early Urban Reform

At the turn of the 20th century, in a variety of ways, the school garden movement challenged ingrained cultural perspectives that had long reflected industrial models of learning. In 1904, the author of a new manual of urban gardening noted, with frustration, that North America needed to follow the Europeans and promote the pedagogy of the urban school garden as a symbol that challenged ideas of ownership. (Klein 1904) Reformers promoted the garden movement as a basis for moral, intellectual and physical development and predicted the rise of school gardens, where "…the moral, intellectual and physical natures of children is so apparent that it is safe to predict that during the next few years a school without a garden of some kind will be an exception" (Ibid., 14)

FIGURE 1. School children in garden, New York City, circa 1912. Library of Congress: Prints & Photographs Division. Photo no. LC-DIG-ggbain-03334.

A new respect for nature and evolution, as separate from human ingenuity, also emerged in some of the writing:

> A boy is no more a miniature man than a larva is a moth. He is the caterpillar stage of a man, the growing stage. Interference with nature in producing a moth results in a weak moth. Interference with nature in building a man results in a weak man. The factor of growth, the factor of evolution, must be taken into consideration in any sane system of education. (Stabbers 1912, 67)

The notion that urban gardens themselves were part of a larger system that improved surrounding community again had its roots in earlier reforms. Documents reveal that some early gardens promoted an innovative focus on the educational and pedagogical value of wild native plants, including flowers, as the basis of teaching and learning. Early school gardens were also advanced in terms of social organization; instructors from the Department of Education traveled to the gardens to instruct garden-centred teacher training. Other advancements were economic and included the cultivation of plants for sale in the cities' markets. When urban school gardens expanded at the turn of the century in the United States, a portion of all school grounds was set aside for the growing of vegetables. (Greene 1910, 320) Community practices in urban school gardens also lead to advancement in government-funded trades and colleges, where gardens were used as a training aid for the children of farmers and laborers.

Despite the various ideological and pedagogical goals of the gardens, the sites themselves typically evolved toward practical purposes of providing food and training. Gardens served national vocational and economic goals, one of which was to stop the flow of unskilled labor to cities by supporting the agricultural potential of the country. (Ibid. 7) Original documents record a society with dual purposes of promoting both organizational efficiency and social responsibility as the basis for learning in urban nature. A garden symbolized the ideal container for both. Some writers of the time were wary of the limits of farm-based learning and saw growing food in gardens as separate from intellectual growth: "School gardens are not intended to create gardeners or farmers, but to afford the growing boy or girl an opportunity for many-sided development." (Ibid. 2) Since the beginning of the movement urban gardens have thus, in some way, always served multiple, civic, educational, economic and vocational purposes.

While recognized by some, the combined pedagogical, cultural and farm value of gardens was far from a centralized goal. The urban school garden was viewed as a container for society's values where the outcomes could be made public and measured with a Victorian's attention to detail. The gardens perfectly suited the tendencies of the time toward a preoccupation with uniform organizational control over nature, space, image and idea. Curriculum design in 1910 describes school gardens in often unimaginative terms, where "The gardens may be collective or individual or both; they may be at the school or the home or both. In all these cases the plants to be grown are much the same and the methods involved in growing them are similar." (Ibid., 2–3) While practices were somewhat dull, the lofty goals usually held moral undertones: "While the purpose of the garden was threefold (educational, industrial, and social) it is only in relation to others that moral conduct or character exists." (Ibid.) In a few rare cases early reformers vividly describe a pathway of potentially varied outcomes of school gardening, from practical civic virtue to magical "fairy tale":

> I did not start a garden simply to grow a few vegetables and flowers. The garden was used as a means to show how willing and anxious children are to work, and to teach them in their work some necessary civic virtues; private care of public property, economy, honesty, application, concentration, self government, civic pride, justice, the dignity of labor, and the love of nature by opening to their minds the little we know of her mysteries, more wonderful than any fairy tale. (cited in Greene 1910, 4)

At the turn of the 20th century fifteen to twenty normal schools and ten to twelve agricultural colleges in the United States were based in school gardens. While the extent of the documentation largely focuses on the creation of school gardens in large Eastern cities, the larger movement, promoted by the Department of Education, spread rapidly across the United States. Los Angeles, California, described as a place of pioneer work in school gardens, was the first West Coast city to develop a curriculum that boosted farming in both rural education and consolidated schools. By 1904, in Oakland, over one thousand children were immersed in garden pedagogy. (Davis 1905)

Normal school educators were seen as risk-taking pioneers who first endorsed the school gardens, in both a practical and pedagogical sense, on a widespread basis. While much of the focus at the Chico Normal School garden, for example, was on tools, timing and teacher control, documents at the time also noted the positive impact of gardens amongst students. From the Chico experiment, reformers were convinced of the experiential value and the central place of gardens in urban schooling, in terms of both practice and pedagogy.

A few universities took an active role in promoting innovative agriculture methods in the community and in promoting the values of community in schooling. In 1910 the University of California provided notable, key financial support to include agriculture in schools with a dedicated plan "....to make school life a copy of the community life, to link the school and the community through the agency of the school garden . . ." (Babcock 1911, 72)

In addition to teacher training and federal government support, private business also gave a direct boost to secular ideology in urban gardens. Around the same time as the normal schools began linking pedagogy with gardening, the president of the National Register Company of Dayton, Ohio, motivated to experiment, began a large-scale garden scheme centered in a low-income neighborhood with a goal of employing urban youth, where the garden workers would take home the produce they grew. The urban garden at Dayton was a turning point where gardens focused on economic progress also purportedly benefited urban youth workers.

As the 20th century progressed, North America's school gardens held competing perspectives around the moral, economic and pedagogical goals for land use and learning. In Europe, school gardens had been recognized for decades as sites that promoted effective secular learning centred on student ability to learn through a discovery of nature. The secular focus on gardens in France was maintained by the government, who controlled the direction for learning in school gardens. French teacher training included practical knowledge of gardens, plants and soil, where the main object was "....not to teach the business of farming but to inspire a love for the country and to develop the natural tendencies of children to become interested in flowers, birds, etc." (Cited in Greene 1910, 9) In Berlin, on a ten-acre central garden in the city, the focus of urban school gardens tended toward Science and the study of Biology or Botany. In Russia, gardens were promoted and funded by the government, and, like France, the Russian government mandated that every publicly funded school maintain a garden for flowers, vegetables, orchard and forest trees, and an apiary. Russian garden-based teacher training was provided during summer courses, courtesy of the government. As was the case later in the United States, garden instructors who traveled to the sites from a central agency ensured control in both design and practice.

In the UK, the government neither funded, nor controlled the direction of school gardens and up until 1900 they were built as supplementary sites, or supported by private philanthropy. Nonetheless, Britain strongly supported the pedagogical connection between learning in gardens and the intermediate, middle grades focus on hands-on learning, a connection still very popular today. In another innovation, the pedagogical evolution of urban nature as an interdisciplinary study can be traced, in part, to the work of Sir Patrick Geddes (1854–1933). In 1902 in Edinburgh, Scotland, Geddes was a pioneer in promoting outdoor learning with the idea of educating the whole person in the context of the urban surroundings. (Palmer 1998) Geddes was an early visionary of an interdisciplinary perspective on environmental education where nature held learning value beyond the single subject area of Science. Such a perspective supported the idea that nature study was not simply a new subject to be included alongside others but a means of progressively transforming all existing subjects to include nature study. (Jewell 1908, 9) There was widespread support in the academic community for Geddes' higher ideals around the integration of learning and urban environment as part of the pedagogical process. Today, the roots of Geddes' ideas can be seen in the promotion of interdisciplinary Environmental learning theories.

In Britain, gardens were not promoted on the basis of economic interest, morality, or duty, but were instead promoted for the pedagogical value they held in teaching students the subtleties of nature from within the processes of nature itself. Agricultural knowledge was integrated into regular middle grades curriculum for purposes of advancing experiential education and other goals. Journals of the era described the curriculum: "Its aim is pedagogical. The utilitarian information given is incidental. The main object of the study is developing the power of observation, the skilful use of the finer muscles of the hands, and by experience through practical lessons in cause and effect." (Greene 1910, 3)

Canada, like Europe, established late 19th century systems of urban development and renewal through agriculture. The movement was lead by Dr. James W. Robertson, a former Commissioner of Agriculture and Dairying, also known as "the agricultural wizard of the north." (Davis 1904) Robertson helped popularize agriculture in schooling and school gardens throughout the Eastern provinces and the Maritimes where he garnered support amongst the school superintendents. Until 1904, Robertson was the director of the Macdonald Fund and President of Macdonald College at Ste. Anne de Bellevue. The Macdonald Fund, funded by Sir William C. Macdonald of Montreal, established the popular Macdonald Schools, which were based in Guelph, Ontario, and promoted through the Ontario Agricultural College; in the early 20th century, Macdonald school gardens held a recognized place in the provincial systems of education. Similar to Europe, and, later, in the United States, Canadian teachers of agriculture were sent from a central body to train teachers and supervise gardens in schools across the country. This system took place in every province in existence in 1904, where teachers from five schools at a time would receive training. Reports from the time suggest that the school garden experiment was received favourably, and with open mindedness, from teachers and administrators alike.

The Canadian model supported the secular European model where the idea of encountering spirituality through nature was included, though not as a central purpose, in school gardens. In early 20th-century Canada the stage was set for innovative, interdisciplinary pedagogical practice and research, which would have been ideally situated in the field of Education. Trends in school gardens, teacher training and vocational training were all converging in a way that may have helped develop a strong foundation for environmental learning in both research and practice. However, here, as elsewhere in North America, declining support for teacher training around agriculture and environment was cited as a reason for the turn toward practical matters of production and yield. Gardens in Canada (such as the famous Macdonald school in the Eastern provinces) were soon promoted as practical sites for the sole promotion of agriculture, and responsibility for pedagogy in the gardens was taken over by government and philanthropic interests. (Greene 1910)

The "New Viewpoint"

Back in the United States, with diminished support from teacher training programmes and academic research, urban reformers themselves began to note the pedagogical benefits of gardens. At the turn of the century, school gardens were large sites where some reformers pitched far-reaching plans that went beyond the growing of vegetables. Fannie Grison Parsons' son Henry, an influential policy maker in New York, promoted his mother's "new viewpoint," which held that a purpose in developing gardens "...was to foster the growth of children as its main object, instead of being principally to teach them how to grow plants." (Parsons 1910, 3) Parsons noted pedagogical possibilities of gardens as a "living laboratory" that went beyond aesthetic appreciation: "In no other laboratory, perhaps, does the child learn so quickly and so clearly that 'knowledge is power.'" (3) In considering the Macdonald school gardens in Canada, Parsons suggested the gardens were a basis for holistic immersion in garden pedagogy:

FIGURE 2. Garden plots, Consolidated School, Guelph, Ontario, Canada 1908. John Woodruff, Library and Archives Canada, Ottawa. C-012112.

> The work of the garden is recognized as a legitimate part of the school program and it is already interwoven with a considerable part of the other studies. The garden is becoming the outer classroom of the school, and its plots are its blackboards. The garden is not an innovation, or an excrescence, or an addendum, or a diversion. It is a happy field of expression, an organic part of the school in which the boys and girls work among growing things and grow themselves....(cited in Greene 1910, 17)

Urban school lots also began to be viewed for their potential to transform students and learning precisely because the conditions for a garden seemed impossible on a concrete schoolyard. Transformation was envisioned in small, realistic ways. As one reformer described: "A paved yard with no sun is enough to dampen the ardor of the most enthusiastic. However, should the space permit, a few of the bricks could be removed along the wall and that space cultivated . . ." (Klein 1904, 112) The potential for widespread, small-scale benefits of school gardens was also recognized as antidotes to the rote nature of most education at the time: "...if we utilize the interest in [the garden]to freshen the wearisome tasks of the schoolroom...The dullest child will brighten as he or she lays out the little plot, figures out the crops, or calculates the gains." (Greene 1910, 36) Some noted the interdisciplinary value of gardens and described how the garden brought newly imagined narratives on interdisciplinary subjects. In a way that still echoes today, the gardens were viewed as sites for holistic learning of nature's systems that also promoted insight into the value of physical work: "It is easy to show how....the life of the whole round world is inter-dependent, or in a child's phraseology, 'hangs together.' No child who has ever loved a garden will despise the farmer for he has learned by experience to respect manual labor, and that brains and hands must work together to bring good crops." (Greene 1910, 37) As manual labor was integrated into mainstream classroom learning, gardens also challenged the strict divisions of urban social class and broadened student perspectives on the value of farms in the city.

Amongst rural and small town school advocates, there was a sense of wariness that their values would be reduced due to "city concerns" that focused on social reforms. Binary notions of rural and urban—the notion that the city was a problem in itself and that rural life represented a perfect ideal—were persistently entrenched in schooling. With a few exceptions, original documents show that the notion of "nature in the city" was rarely promoted. If nature existed at all in the city, it was typically viewed as a rural ideal of the popular imagination or as a means of re-creating paradise on earth.

However, overall, early documents do express optimism around the grounded, progressive outcomes of learning in school gardens. While an emphasis on social control was tempered by secular, economic and vocational interests, there were always those who hinted at larger ideals, where gardening helped students find solutions they could apply to their own life:

> Why should you give your pupils the benefit of a school garden? Because it brings living principles home to the children, and school is living—not a preparation for life. Because it enables the children to solve for themselves, under the law of necessity, some of the most difficult problems which the school course has to offer. Because the garden supplies ideal conditions for cultivating the hand and the heart as well as the head. (Palmer cited in Greene 1910, p. 37)

The idea of appealing to students' minds, hearts, and experience, as well as focusing on the earth, and the dignity of manual labor all within the school day, was a pedagogical advancement in the educational standards of any era. At the time, both urban farming and the inclusion of the natural environment was far from mainstream education. Reforms focused on social aid for urban youth (whereas rural areas focused on practical applications that would yield economic benefits). Several reformers noted how a garden comprehensively improved agricultural processes and aesthetic appreciation from a local site that then helped transform understanding of the surrounding city.

School gardens offered students, teachers, and reformers an important way of thinking about learning that emerged from the relationship between humans and the natural world. At times the pedagogical possibilities were described in terms of student-led learning, which brought even further, rare possibilities of learning in nature. The work of gardens was newly legitimate and rooted in experiential learning principles where student understanding progressed organically from hands-on problem solving to the blossoming of new ideas. As one writer of the time explained: "They learned to do by doing; proceeded from the known to the unknown. Their curiosity was satisfied when their interest was at white heat, nothing ever being forced upon them. They were instructed when the time was opportune and a desire for knowledge had been awakened. (Klein 1904, 59)

Learning around local land, instilling pride in urban dwellers, inspiring grass roots practice—such ideals form the basis for urban learning in gardens. Reminding learners of the earth and of their own unique learning environments in the city was a first step in garden pedagogy and an important task for educators that persists to this day. Today, the words of the early reformers sound similar to current perspectives on environmental education that promote experiential learning, revitalization of curriculum, the empowerment of students, the promotion of nature as central to the learning process, and the power of local learning where an urban school garden leads a learner "to see that he can do something worthwhile right where he lives." (Davis 1905, 11) In 1909, in a modern-sounding description of environmental learning, Clarence H. Weed discussed the work of urban youth building their own gardens in their tenements, and describes multi-generational, multi-racial groups of mixed gender all sharing knowledge of farming practices from their elder parents and grandparents.

Before WWI, independent reformers urged school administrators to modernize and develop school gardens as a way of addressing issues of underutilization of urban green space as well as vocational training of urban youth for trades in farming. The recognition for urban land as farm land was, again, a progressive notion and reformers began writing about the more practical purposes of urban sustainability with a social goal of boosting the income and self-sufficiency of urban wage earners via enhanced use of plots: "With a common knowledge of the principles and possibilities of crop production, the wage earner of the future will not need to measure his income solely by the size of his pay envelope. He will consider, also, the productive capabilities of his garden plot and the extent to which it will reduce the cost of living . . ." (Jarvis 1916, 9)

Since youth in cities had to work at an early age, usually in mills and factories, urban school gardens also provided access to vocations that promoted self-sufficiency based upon land use. As one reformer wrote: "It is not necessary to bring in statistics to show the great inequality in education." (Jarvis 1916, 10) The garden was seen as a way to solve blatant urban problems around high dropout rates and the use of youth workers in factories and industry. Garden also helped ease dissatisfaction with schooling amongst parents who, in another echo of today, believed that schools did not provide enough practical, relevant learning opportunities specialized to meet the needs of urban students. Gardens seemed to provide a visible solution to the general problem of lack of vocational opportunity for urban youth. As a hint of pedagogical innovation and experiential learning skills in a garden it was noted that garden skills: "…are acquired, not taught." (Ibid.) At this time, urban school garden pedagogy emerged around the principles of "hand, head, heart, health" (Ibid., 16), the basis of the American "4H" movement, a group that is still very involved in agriculture in the classroom today.

For some reformers just prior to WWI the concept of an urban school garden was not simply meant to transfer the values of rural life to cities. Support began for the role of urban gardens themselves in alleviating the problems of the crowded city. (Jarvis 1916, 23) As urban pedagogy the garden was recognized for providing a space for learning based on trial and error as a welcome part of the learning processes, as one educator noted: "Boys and girls, despite careful direction, are sure to make some mistakes, but these constitute a very important part of their training" (Ibid., 52) Such modern-sounding statements from urban reformers and garden supporters represented a break from Victorian era education, while also providing an innovative, student-centred approach to learning. As new pedagogy, the garden became a catalyst for innovation and change where learning itself was changed by the garden.

While the gardens may or may not have started with a pedagogical focus the practical cultivation of the spaces themselves held far-reaching effects where "even the desire to start a garden lead to innovation—that the teacher would learn alongside the students, instead of 'having it all worked out ahead of time.'" (Babcock 1915, 338) The setting itself changed how the teachers taught, with a goal toward engaged students learning alongside teachers. Writers of the time noted that "Even an excellent text cannot supply the lack of actual experience to the teacher....a teacher can study a new thing with the pupils in the spirit of co-partnership and while so doing secure a new hold on the pupils and fresh interest in the work." (Ibid.)

A side-benefit of practical reform work was a renewed focus on local urban economies and democratic communities. Since urban farming was a relatively low investment for vast yield, and based on manual work instead of on market speculation, the work also supported another progressive notion: a sustainable, local economy. Urban gardens represented a practical market economy rooted in the knowledge, skill and motivation of individuals and small communities in self-organized activity. Almost a century later, such an ideology would be mirrored in notions of ecological justice and "Earth Democracy" where "ecological and economic identities are linked to roots in a place and community."(Shiva 2006, 81)

As a form of local, grass roots community action, an urban school garden still represents an effective, sustainable model for experiential, localized learning. As Booker T. Washington said about the value of learning from within nature:

> I have gotten a large part of my education from actual contact with things, rather than through the medium of books.....I like to watch plants grow...For the same reason, I like to deal with things, as far as possible, at first hand, in the way that the carpenter deals with wood, the blacksmith with iron, and the farmer with the earth. (Washington 1911, 421)

How did the urban garden movement progress? At the turn of the last century, urban school gardens were a large national, and international movement, engaging thousands of students on thousands of acres, promoted by industry, educators, writers, teachers, government, parents and local groups. Today, large scale gardens built on the East Coast have vanished, in most cases leaving little physical evidence they ever existed. One explanation is that because gardens were linked to local community interests and to an interdisciplinary, constantly changing roster of educational, political and civic bodies, there was no central academic research body from which to research and support the movement.

By 1916, even with community support, without the backing from educational authorities and policy makers, reformers badly needed the support of teacher training to provide a pedagogical basis and ongoing research for their work. Reformers grew more and more critical of educators who supported school gardens as a political goal unto itself or as a means of public relations. In a way that also sounds very familiar today, due to shrinking budgets, hectic schedules and large classes, teachers simply did not have either the time, foundational teacher training or professional support to engage the deeper pedagogical principles of school gardening that would help establish and promote the movement. After WWI, the main research group that supported school gardens, the Nature-Study-Movement (at Cornell University) was defunct. (Trelstad 1997)

In 1914 school gardens were 14% of the entire federal budget for education. In 1918, the U.S. entered WWI, and the School Garden Army assembled to address demands for local, domestic vegetable production. Support for school gardens thus transformed into support for community Victory Gardens. After WWI, the once expansive garden movement lost speed. During and post-WWI school gardens served a purpose; when the war was over, gardens seemed unnecessary. At this time, summer camps also became popular and fulfilled the need to connect urban youth to the health benefits of nature. The idea of recreation as a pursuit unto itself also rose in popularity. During the 1920s factories moved to the edge of cities and urban planning somewhat improved the grittiness of daily urban life. The rise of the suburbs stimulated peoples' interest in private home gardens over common space as societal focus shifted from shared community and neighborhood, to family. Without a "school for school gardeners" (Benedict 1913, 262) it was widely acknowledged that a lack of both teachers and support from school authorities combined with lagging support for agriculture-based curriculum, or basic farm practices, the pedagogical component of the movement lost pace.

FIGURE 3. *Join the United States School Garden Army* Poster. 1918. Photo by Edward Penfield. 1918. Library of Congress: Prints & Photographs Division, photo no. LC-USZC4–10321.

Gardens survived in some larger cities (for example, in Philadelphia and Cleveland) where reform was long established and where innovative educational value was as highly valued as practical (i.e., food) concerns. Even as the small garden pedagogy movement withered in some quarters, some cities doubled their budgets in order to support teaching, pedagogy, nature-study and gardening. Overall, gardens located in cities that tended to value the educational and economic benefits of gardens along with progressive, civic reform, also tended to last. By the 1930s, poverty was widespread, and therefore there were fewer assumptions made about the role of gardens in improving the lives of urban dwellers since many universally suffered under the collapse of the capitalist economy. (Lawson 2005)

During WWII, households and communities again built Victory Gardens for multiple purposes. Urban gardens were now called War Gardens and much available land went toward growing food. Military themes and metaphors heightened the association between gardening and patriotism. Gardens were described as "munitions plants and gardeners as soldiers of the soil" (Ibid., 139) Immediately after WWII global food shortages again emphasized the purpose of school gardens towards necessity. School, community and governing networks worked together; new collaborations were made.

When the wars and the immediate food crisis ended, school gardens, in both practice and pedagogy, withered. Apart from the 1960s,70s and late 1990s, most of the rest of the 20th century had a very limited, if not abandoned, perspective on the socio-economic, pedagogical and cultural benefits of working the land. In planning cities in close connection with nature, "Professional wisdom held that….either too little nature existed downtown or too little was known about it for such an approach to have practical results." (Spirn 1984, 3) In the 1970s, community and school gardens rose among other projects related to the decades' social movement. Such activism arose out of a declining urban population; gardens were developed in abandoned, post-industrial sites as a metaphor for universal access and an improved standard of living. For some time, focus shifted to community gardens. Out of all of the school gardens that flourished at the turn of the century the Cleveland school garden program had the greatest longevity, running from 1904–1973. In the early 1970s, school gardens at ten schools in this program became community gardens. By the 1980s, it was obvious that the demise of the movement, both in pedagogy and practice, rested on political perception of space and on the politics of land use. The 1980s was an extension of the post-war "war" on school garden space. Ultimately, throughout a good part of the century, school gardens had been an ideal focus for supporting land, learning, urban renewal and pedagogical innovation. Without political support, research, teacher training, urban reform or a food crisis to ignite interest in gardens, the economy shifted focus elsewhere.

Then and Now: A Current Case Study

Over a time period of one hundred years, the stories of urban school gardens show cycles of ideals in action and reveal just how precarious, hopeful, and, heart-breaking the work of urban school gardens can be. The history shows stories filled with dedication, drama, and difficulty, as clashes over different value systems and ideas of land ownership and use are manifested on sometimes the smallest urban plot. Until the 1970s, however, few urban school gardeners wrote extensively about their firsthand experiences of the deeper conflicts. The following case study is unique as it provides an insider's candid perspective that contributes to the evolution of school garden culture as it developed through the 20th century.

In 1971, Herb Wong and Robin Moore (1997) began a ten-year action research project called The Environmental Yard (ESY) where they redeveloped an asphalt urban schoolyard into an educational resource and community open space. The site was at Washington Elementary School and the K-3 University Lab School in Berkeley, California. In the 1970s, links between social justice and a UN movement for children's rights laid the groundwork for learning that combined social and ecological ideals aimed at improving education. The authors compare the development of their garden as "a drama" with many involved; some stood in the path of progress, while others moved forward, struggled, and painfully watched their gardens removed. The teachers compared learning in a garden to "a pedagogy of movement that is like a city itself." (xviii) And, as was the case throughout history, the challenges made way for foundational learning: "Conflict was inevitable and reflected the heart of the adventure: taking risks, overcoming barriers, plotting strategies, and helping nature to heal itself. On the other hand, the yard offered adults a hidden treasure: a magical doorway for the imagination, through which every child could peacefully co-exist with nature." (xviii)

It was obvious from the start that the controversy was not simply based on land or contested public space. The garden represented an alternative and a new way of teaching that challenged the status quo: "Without a doubt, the Yard illustrated an effective alternative—one based on respect for children and a belief that educational success must be supported by improvements in their quality of life at school." (xvii) The "drama" in the schoolyard created a learning situation where "....dealing with people's prejudices and personalities took more time and energy than planting trees and digging out ponds. Some of the most heated discussions were prompted by the proposal to reintroduce the natural environment into the school site where all the local flora and fauna had been eliminated." (12) Somehow, the restoration of the public space, the return to nature, and the removal of the man-made elements of the schoolyard sparked major controversy. Even the idea of the garden, from the start, revealed fears: fear of vandalism, of injuries, of students hiding, of falling out of trees, of falling into water. For Moore and Wong, much of the learning in the garden emerged from the controversy and from the struggle to introduce organic elements of nature into the typically sterile world of classroom teaching. Their story shows how the function of the urban garden was to grow plants and to expose, and examine, ingrained biases that thwarted the inclusion of the natural environment into daily schooling.

The authors discovered their own ideals by learning about the systems of education that restricted learning and repressed the truth in stories: "To sing only praise while hiding an often nasty struggle would be to falsify, and thus diminish, the actual situation....Often, the liberated imagination of children, the wondrous reclaiming power of nature, and our belief that the outdoors could be a powerful supporter of child development were all that we had to keep us going." (13) In the end, the garden brought more unanswered questions, than easy conclusions as the authors realized that their work was spontaneously changing urban education by revitalizing the urban landscape. As they note: "...the Yard was far more complex and dynamic than we had ever imagined." (23)

The authors describe how their original motivation was inspired by historical photos of children working in turn-of-the-century gardens. They were curious to know why such a large-scale movement had faded, so they set about developing their own garden. They deduced that educational priorities had shifted since the first gardens were established in North America, and that this had lead to the demise of the work. Their goal for the garden was therefore to:

> reintroduce the natural environment as an interactive educational resource, with gardening as a leading strategy. If schools had had gardens so long ago, it did not seem especially radical or impractical to reintroduce the idea fifty years later. Gardening provided an easy way for teachers to initiate interdisciplinary environmental education...(25)

Like many urban school gardeners, the teachers started with the logistics of their gardens, the measurement of the boxes, the soil content, the hard work required to transform a physical space; yet like so many others, practical work gave way to other, unexpected realizations. Moore and Wong talk about how the space even changed how students spoke, how they expressed themselves in new ways in the garden, and, how the students found language right in the landscape:

> Outdoors, language was everywhere. 'It's like hiking through a dictionary,' enthused one teacher. You can walk through the Yard and see the words all around you, feel them coming at you, and watch the children use them in every moment in their play. (125)

Despite the progress made in urban renewal and learning, the garden went the way of most public learning gardens. In the end, the school was renovated (1995–96) to meet new earthquake standards and The Yard was replaced with a rectangle of grass and basketball courts. In a disheartened postscript, the authors state:

> As the yard's life history comes almost full circle, one message is clear. The creation of high quality outdoor school environments will require a change in both values and governance in K-12 education. (251)

The story of the Environmental Schoolyard exemplifies 20th-century perspectives on human interaction with urban land, and the influence of constantly shifting cultural values in that interaction. As the authors realized through firsthand, painful experience, values, ideology and school governance always surrounds the work of land stewardship. Whereas firsthand school garden narratives of past centuries tend to be absent from the literature, the seminal story of Moore and Wong shows how a garden that may not physically survive still helps the larger movement, in practice, pedagogy, and ideology, live on.

The Future: Garden as Culture and Story

> We know that there were ceremonial gardens in several ancient cities of the Inca and Aztec empires, and that there were thousands of floating gardens on the lake of Texcoco, and yet we know very little about them, their uses, their significance for the lives of the gardeners and the communities where they were located. We would like to know whether they...contributed not only a meaningful grounding of self but also 'structures of feeling' enabling people to share in some emotional dimensions of life .(Conan & Quilter 2008, 5)

From the 1890s onwards, urban life declined as populations moved to the suburbs and countryside. Air and water quality worsened, unemployment rose, and the popular notion of rural idealism was born; a garden in the city was viewed as a smaller, remote version of paradise. In school gardens, support for the commons has historically met with industrial notions of land use, ownership and learning. History and firsthand accounts show how gardens nurture new, unexpected, creative forms of expression and learning. Then as now, real innovation and courage is required to make visible certain ideals around the use of shared physical place. The urban school garden remains a site for a revolution of new learning and ideas. And the articulation of the value of gardens is and always has been important as gardens struggle for recognition in schooling and in campuses of higher learning.

A pedagogical and cultural history of urban school gardens illustrates a few broad themes and practices that remain, and remain innovative, to this day. One progressive notion central to urban garden-based learning both then and now is the reclamation of unused asphalt and playing fields in schools to create shared places for green learning. A repeated theme is the urgency to revitalize and re-populate regions through urban agricultural training. Today, as seen in the case of Detroit and other depopulated urban areas, the focus has recently shifted from rural to urban re-population.

Another change over the years relates to dramatic shifts in perception of use of public space. For example, in the late 19th century use of public land for appreciation and observation was considered passive by reformers of later eras. (Cranz 1982) Victorian ideals tended toward social control, manifested in schooling and garden design; paradoxically, they strongly supported immersion and appreciation as active pursuits. Later, the simple observation of nature itself was considered passive non-action and public perception shifted where highly visible land use held value. In some ways, what was lost since the Victorian era was the recognition of the strong, subtle, and typically invisible connections between learners and land. While educators now struggle to link learners to the natural world in highly visible ways, one of the most progressive aspects of the gardens has been their ability to convince learners to share, observe and appreciate the land in a way that challenges ingrained biases toward ownership.

One of the most dramatic changes in the past one hundred years is the emergence of urban garden culture. Ultimately, the pedagogy of the urban school garden did not evolve as a linear narrative representing a series of progressive steps where each step successively bypassed the tradition before it. In fact, in Western society the movement could be viewed as radical since it includes traditional practice at the core. For example, aspects of the current garden movement are rooted in ideas and methods in place over one hundred years ago which, even at the time, were modeled on traditional farming techniques. One constant of the sustainability movement is, paradoxically, the reconnection with tradition, history and community.

The history of human interaction with gardens is a story also linked to changing social expectations and notions of progress in how humans interact with place. For example, the rise of the sustainability movement as it now coincides with global environmental change is again sometimes spoken about in terms of social, moral urgency and values. This time, however, the theme includes the idea of a land ethic where the earth, in a state of need, is valued and supported for its own ability to sustain itself. In one hundred years the focus has also shifted from "saving" urban youth to engaging urban youth in community and the land in a way that sustains human life in the context of environment.

Urban school gardens are not new, and gaps in the story remain; many so-called advances simply vanished. For example, at the turn of the last century, teacher training was in fact closely linked to farming. Now, even as farm land disappears to development and food security moves to the top of the agenda, the notion of integrated professional training in teaching and farming is virtually non-existent. (Gaylie 2009) Another major gap is the lack of first-person narratives and pedagogical research from teachers, students and others in urban gardens. The lack of recorded, personal story around the sites is also telling of institutional perspectives, ideologies and values around shared urban land.

Over time, certain elements have been a constant. While the strongest support shifted from formal to informal institutions and community groups, the legacy of the gardens that persists across time has involved: the vision and courage of individuals; the shared vision and work of community; the development of interconnected networks of support; increased support for the movement in the context of community and popular imagination; strong community and social organization to bring forward the value of small projects and large perspectives; and the ability of individuals and groups to learn from previous work and ideas.

Today, over one hundred years after school gardens were first created in North America, gardens in schools have evolved into participatory events with students involved in the design, planting, maintenance and, more and more frequently, in valuing the experience of nature in their own urban neighborhoods. What is known now, from a pedagogical perspective, is that a garden allows learners to connect to the land in a direct, experiential learning process central to their own learning and community. Today, urban learners gain experience in practice and pedagogy, through agricultural training, and through projects in ecological and social justice that connect their local spaces to the world.

Beyond the notion of a school garden as teaching tool, such sites are a symbol of cultural evolution in how Western societies perceive place itself. Connecting urban learners to the land, to their inner-city dwellings, helps create stories that bring to life the pride and sense of belonging that arrives when land is raised as a cultural symbol of experience. In recorded history, the cultural conceptualization of space has typically lagged in Western, capitalist perspectives. The industrial world brought a geometric utilitarian understanding, and a square garden, to schooling. Despite all, the garden has brought learning and pedagogy above and beyond human expectation and control. In future, garden pedagogy will be increasingly grounded by environmental design where space is viewed as "a pre-condition of our relationship to the world" that "lies beyond any intellectual control, beyond any critical inquiry into the transformation of symbolic qualities of space...." (Conan & Quilter 2008, 3)

Innovative research and practice in the future will focus on the cultural value of land and in non-Western perspectives that acknowledge a deeper respect for the land, where "The better we attend to our locationality, the better we are able to understand what is at stake in our lives." (Donald 2009, 1) New practices and pedagogies will look to other non-Western, non-industrial perspectives that have elevated land to cultural practice, including, for example, the work of Cuban revolutionary, educator and philosopher Jose Marti (1853–1895). In the late 19th century, Marti, also a poet, was involved in urban education, agriculture, language, literature, and teacher education. In New York City, he was a high school English teacher, an interdisciplinary educator, and a critic of a narrow formalist view of schooling that removed learning from practical experience. As a social reform visionary, Marti preceded Dewey's ideas about experiential education, Shiva's ideals around environment and social justice, and Orr's conception of environmental education. Over 100 years before the popularity of the "100 Mile Diet" (Smith & MacKinnon 2007), the "100 Mile Curriculum" (Giesbrecht 2008) or "Farm to School" movements in North America, Marti widely supported the common sense idea of eating and learning locally as a means of supporting, sustaining and learning about the practice and cultural value of local agriculture. As he stated in 1884: "Men will always need the products of Nature. And since every region produces only certain products, its trade will always be kept alive, thus assuring wealth and freedom from want for all peoples." (Marti 1884, 38) As a model practitioner/philosopher for the current movement, Marti lent his views on agriculture linked to farming practice and to pedagogy; he was a poet at the same time he responded to food crises amongst the urban poor. His work in pedagogy and farming helped shape the understanding that the practice and meaning of gardens is "...not be explained in formulas on the blackboard, but rather in the layers of the earth itself." (Marti [1884] cited in Foner 1979, 13)

His philosophies around urban learning in and through agricultural land apply to pedagogies in urban school gardens, where physical labor brings empowerment and grounded transformation in both land and learner. He uniquely developed a culture of urban pedagogy, promoting education reform beyond goals of social or behavioral control. Marti's ideas uniquely relate to the pride, freedom, and knowledge of manual labor, and his legacy continues to influence Cuba's urban school gardens today, for example in the Youth Pioneres (Youth Pioneers) after-school programme that involves nearly 2 million students who work in, and learn about, the environment through gardening (Subramaniam 2002, 56) Other exemplary garden projects include Los Niños in Mexico and the Garden of Peace in Brazil, both of which promote democratic values of fairness, equality and respect.

Beyond the notion of a school garden as teaching tool, such sites are a symbol of cultural evolution in how Western societies perceive place itself. Connecting urban learners to the land, to their inner-city dwellings, helps create stories that bring to life the pride and sense of belonging that arrives when land is raised as a cultural symbol of experience. In recorded history, the cultural conceptualization of space has typically lagged in Western, capitalist perspectives. The industrial world brought a geometric utilitarian understanding, and a square garden, to schooling. Despite all, the garden has brought learning and pedagogy above and beyond human expectation and control. In the future, garden pedagogy will be increasingly grounded by environmental design where space is viewed as "a pre-condition of our relationship to the world" that "lies beyond any intellectual control, beyond any critical inquiry into the transformation of symbolic qualities of space...." (Conan & Quilter 2008, 3)

Across time, worldviews, culture and economies, urban gardens represent the history, and the future, of the relationship between land and people. An urban school garden is a history of contested space, courage, renewal, and hope. A garden as commons potentially brings new lessons for inner-city students and educators struggling to learn and teach in industrial societies where connections between learners, learning and land are increasingly remote. The cultivation of an urban garden invites students to re-imagine land in the context of its history, region, culture, and story. Whether the urban school garden represents land, design, politics, food, reform, or pedagogy, firsthand stories from those in the field bring understanding of the practical, pedagogical, and potentially transformational benefits of such sites.

· 3 ·

SOUTHERN CALIFORNIA

Gardens in the Desert

What happens when you deny that your river is a river?
(Price 2005, 229)

In the late 19th and early 20th centuries Southern California represented a garden paradise where nature was cultivated to perfection. Advertising campaigns boosted the population by promoting the wheat boom in the central valley and new strains of citrus in the South. The region invited a new life for those in larger, older industrial cities on the East Coast. However, the selling of the "the California dream" soon led to a rise in population, land development and a return to an industrial economy (i.e., oil) in order to address the needs of the high populations attracted to the area for its scenic beauty. The evolution of urban Southern California, particularly the areas around Los Angeles and the inland desert, very much influenced how the natural environment was perceived by the larger population in the region.

What is the cultural role of the urban school garden in such a setting? Historical evidence combined with ground level perspectives at an urban school describes the challenge and paradox of Southern California's approach to the land; in the midst of a natural paradise, lies urban development and ecological degradation. For this reason school gardens in Southern California are hard won and bring important new perspectives that come through community, labour and an urgent need to re-connect with local environment. The school gardens in this region have an important role: to challenge ingrained biases around use of the natural environment and to bring inventive approaches to sustainability that will aid in the development of a new learning culture. Here, environmental projects invite a renewed interpretation of human interaction with the land beyond notions of ownership or re-creations of paradise.

As humans, nature and cities closely compete, Southern California is also a cultural microcosm that potentially bring new, progressive understanding around use of common space. Here, a school garden represents a small, effective solution to problems faced in small and large communities around the world. For example, in at-risk areas, where natural resources are under intense stress, how do school gardens introduce a practical cultural understanding of land? How is the culture of the urban school garden developed by students and teachers in the region? How do school gardens in the area in fact lead the way as symbols of survival for other parts of the planet facing similar problems? In the evolution of environmental theory and garden pedagogy Southern California is in many ways ideal for prompting new and progressive ideas about urban land and learning. As one author notes: "LA is the ideal place in America to completely rethink what it means to write about nature." (Price 2005, 221)

Garden Paradise

Southern California exemplifies the way that politics, economics and advertising drive the direction and planned use for land and natural resources in urban areas. California's high energy consumption, fossil-fuel economy and rampant urban/suburban development bring challenges to an area already strapped for water and renewable forms of energy. Even in a climate and region ideally suited to farming, gardening and environmental awareness, the natural area itself has been largely ignored in decision making. Southern California's urban nature is itself a paradox. In the early 20th century, high populations were attracted to the region's lush landscape which soon required an industrial manufacturing economy that would both support the high population and dramatically alter the landscape. The social history and geography of the place reflect extremes where the environment was constantly sacrificed for economic gain. A history of the area reveals how, in promotion of the place, place itself was largely forgotten.

In Southern California from the 1870s to 1920s cultural identity was linked to the environment at the same time the environmental was neither protected nor seen as central to survival. For early Mormon settlers identity and economy was associated with the natural world; plants were associated with both economic wealth and religious notions of paradise. For the Mormons in California "turning landscape into an economic engine" was seen as part of "the magic that growth machines perform." (Sackman 2005 p. 247) The hopes of a garden ecology in fact became a garden economy where trees and flowers were only valued in terms of the wealth they produced; in turn, one's ability to extract wealth from the natural world meant a rise in social status. In Los Angeles, garden growth, economic growth and notions of appreciation of nature were "linked in one vast material dream." (Ibid.)

In the Southern desert in and around Los Angeles, orange groves and oil rigs stood beside schools; the cultures of industry and education, farm and city were intertwined, and grew together. The bio-diversity of the LA flatlands, once a treeless prairie, disappeared after Western European settlement. Higher populations and development threatened the native species of the area as livestock and invasive weeds moved in. (Schiffman 2005) The ideal notion of Californian scenic perfection was, by the 1960s, integrated with toxic industry as neighbourhoods were inundated with oil derricks, and the sights, smells and sounds of industry. (Johnson 2005, 81)

At the same time that East Coast urban land reformers were heralded for their work in supporting public access to nature in industrial areas, in Los Angeles, a "garden city," the industrial capitalist economy led by land developers and politicians still severely restricted the connection between people and nature. Some urban theorists (particularly Mike Davis in *Ecology of Fear*) speculate that had Frederick Law Olmstead Jr. and Harlan Bartholomew in fact implemented their parks plan for Los Angeles, the area would have had a dramatically different landscape with an enlarged commons, limited urban sprawl, and preservation of important natural eco-systems such as the L.A. River.

Los Angeles provides an example of how land use and perhaps more importantly, the way people think about land, dramatically shapes a city, a culture, and an education system, over time. In a social democracy, people feel mobilized to preserve shared spaces that contribute to the overall good of a community. The opposite is also true: where shared spaces are non-existent, people feel less inclined to participate in social democracy.

As the population rose and once fertile farm land was paved and filled with highways and fast food outlets, various community groups also mobilized to help one another create a self-sustaining urban food system. By the 1980s and 90s, food security became an urgent issue and in terms of public response, Los Angeles was a leader. Many participated in food banks and public awareness rose about the lack of land for farming. Thus, out of need, progressive, grass roots organizations with innovative, small-scale solutions received a great deal of community support. One community-based group, the South Central Farm, started in inner city Los Angeles in 1992. The site was supported by local, immigrant families or those with restricted incomes, as a place to grow food. A spokesperson for South Central Farmer Feeding Families Coalition, Tezozomac, spoke about the value of the farm:

> In an area like Vernon....the fifth lowest income in the city....after you pay rent maybe you have two hundred dollars for the rest of the month to feed your family. What this project has done is to give people access to high quality produce. People grow their own food....and they can cut their bills by almost a half. We have over 150 varieties of plants that are grown for consumables and medicinal purposes. (Tezozomac 2006)

In 2003, a local developer wanted to purchase the South Central farm land to develop a warehouse. The development plan attracted protestors, Hollywood celebrities, environmental activists, and documentary film makers, who fought to save the farm. Ultimately, the city was unable to secure the land and the South Central farm moved to a new location. The event highlights the contested space of an inner city and reveals how collective farming in a highly populated urban area, where people simply want to feed their families, can grow into a large-scale political action. First-person accounts help describe the importance of shared local space for urban survival:

> I never thought that growing food would make you ....so political. we're just trying to feed our families ....South Central Farmers Feeding Families. That's all we're trying to do is feed families. We didn't come out here to be political. But we've done this out of sheer necessity in order to survive, we have no other choice. (Tezozomac, 2006)

As shared, urban green space, community and school gardens in Southern California have evolved as realistic, innovative, and hard-won versions of garden paradise. The gardens are also a reminder that connection to the land is essential to survival. As sources of food, gardens in this area reconceptualize the notion of an *easy* paradise into practice and ideas that involves labor, knowledge and community as the basis of survival.

Early School Garden Pedagogy

> In LA, we've lost track of an entire river.... We cannot find LA's major waterway—which runs under ten freeways through the heart of LA County. A fifty-one mile river in plain sight: lost. (Price 2005, 229)

At the turn of the century, while PR companies and politicians peddled the natural beauty of Southern California, attracting the millions of people who would one day inhabit the place, educator B.M. Davis wrote a manual called "Nature Study in the Los Angeles State Normal" in the journal *The Los Angeles Normal Exponent*. (1901) Much of Davis's perspectives on learning in gardens grew out of the Los Angeles and Chico State Normal Schools. The manual was intended to provide instructions on how to observe, approach and learn from nature. While farm land was eyed for development and Los Angeles waterways were lost to both cultural consciousness and socio-economic planning, Davis was keenly interested in the processes and systems of nature as they related to learning. He based his garden pedagogy on the notion that: "ALL LIVING ORGANISMS ARE AT WORK AND ALL ARE DOING (each in its own way)." (Davis 1901, 7) He described how "Nature study has largely to do with the relations of things" (Davis 1901, 8) and went on to explain *Facts to be Kept in Mind in Handling Material in Observation or Recitation* while teaching and considering such relationships:

1. To the whole of which the object is a part.
2. To natural environment.
3. To past and future.
4. To other individuals, similar or otherwise.
5. To cause and effect.
6. To man.
7. To Creator. Natural laws.
8. To other school work. (Davis 1901, 8)

At the end of the list, Davis wrote: "These suggestions are expected to help teachers in preparing lesson-plans by keeping the great facts of animals and plants before them." In Southern California, the pedagogy of gardens was, from the start, linked to experiential learning and to an understanding of ecological systems. Davis's progressive, practical advancement in ideas from garden to pedagogy to experiential learning shows an acknowledgment for land ethics and respect for the environment: "To rear a plant successfully, to be responsible for its life, to protect it and minister to its needs, to become thus vitally connected with it. . ." (Davis 1901, 42). Over one hundred years ago, Davis recognized the interconnectedness of both land and learning. Course outlines for Kindergarten through Grade 8 were open-ended and listed ways to connect learners to nature:

- Garden work every day.
- Animals and plants in the rooms for observations in connection with stories.
- Trips to parks.
- Children encouraged to bring in anything of interest in nature.
- Garden work as often as necessary to care for the plants started in the fall.
- Harvesting of corn, beans, peas, etc.
- Field lessons after the rains.
- Time taken for observation of any objects of interest at any time (10).

Specific instructions for using local nature as a basis for learning were specifically noted under the section "Where to Go":

> Any quiet place where water has been standing for some time. The pools along the Los Angeles river may be reached by taking the Daly street car at the Buena Vista street bridge (13).

One of the most valuable entries in this document is a rare, first person narrative, written by one of the children learning at the Los Angeles Normal School, published in *Junior Naturalist Monthly*:

> I dropped a seed into the earth. It grew....No one could tell me why it grew nor how. It had secrets all its own....yet this plant was my friend. It faded when I withheld the light, it withered when I neglected to give it water, it flourished when I supplied its simple needs. One week I went away on a vacation, and when I returned the plant was dead; and I missed it. (Bailey, cited in Palmer 1910, 3)

The excerpt reveals the student's close personal connection with the environment. In a time when schooling was largely based on a non-localized methods of instruction, candid and spontaneous recognition of local nature as part of a students' own learning patterns was rare. In 1910 in Southern California there were calls for the inclusion of school gardens and for greater school teacher involvement in agricultural-based learning on a widespread basis.

> Having conducted the work with several hundred students, and watched as many grade pupils at their garden work, the writer is firmly convinced of its value, and earnestly

> urges every school, which has available ground, and a teacher willing to undertake the work, to profit by the opportunities offered by the growing of plants (Davis, 10).

Almost one hundred years later, the words echo California State Superintendent Delaine Easton (1995):

> Gardens offer dynamic settings in which to integrate every discipline including science and maths, language arts, history and social studies, and art; Young people can experience deeper understandings of natural systems and become better stewards of the earth....(cited in Desmond, Grieshop & Subramaniam 2004, 26)

Sample Gardens

The sample school gardens in this study are located approximately 50 miles east of Los Angeles, in the Inland Empire. I made short visits to three schools (one elementary school in San Bernardino, one middle school in Fontana, one in North Redlands) and a longer visit at Kimbark Elementary in San Bernardino. The cities east of Los Angeles in the Inland Empire were once considered suburban. Over the past several decades large parts of this urban area mirrored several issues associated with an inner city: over-crowding, criminal activity, lack of adequate housing, lack of legitimate employment opportunities, youth and families at-risk. Air quality is often poor, caused by industry, and the area has some of the highest volume automobile traffic on the continent. In this urban area, once a broad, thriving agricultural valley filled with farms and fruit trees, fast-food outlets and highways abound. Access to affordable local fresh food is a constant challenge.

And yet, in this intensely urbanized region, school gardens grow. I visited hard-working schools that sit like miracles in the desert. I met teachers who work hard on shoe-string budgets, or no funding at all, to make a difference in their small corners of the world. While the specific motivations for every school garden varies from place to place, all of the teachers involved in this work are dedicated to connecting learners to the natural environment. And, the teachers all see that connection as central to an urban student's survival.

Elementary School, San Bernardino, California

The garden at an elementary school in San Bernardino sits in a corner of the asphalt schoolyard surrounded by a chain link fence. The garden, though very small, contains surprises and a diverse array of plants and activities. The site itself is a miracle of garden design.

Grade two teacher Linda Gregory built the garden on the paved playground in the year 2000. The garden contains, or has contained:

- a flagstone patio with cement blocks for seating in an outdoor classroom area
- a worm bin, a compost bin and a small garden shed
- recycled large truck tires for planters for cactus
- recycled smaller tires for smaller plants (Amazingly, the students turned the tires inside out and painted them)
- simple raised beds
- a rabbit hutch
- a butterfly habitat garden, which included native plants (important in the California desert region) and flowers that were host plants for different butterflies (i.e., passion vines)
- space in planters for other teachers to use
- a salsa garden (i.e., chilies, tomatoes, corn, cilantro)
- herb garden
- an American Eagle topiary (after September 11, 2001)
- radishes (apparently a delight for 2nd grade students)
- tulips for a "Journey North" Garden (See Activities section in this chapter)
- a water garden, containing important life forms
- a decomposing log to show students the life cycles

Many features in the garden were built from previously discarded items (tires, wood, fencing).

Lynda's advice for would-be urban school gardeners is to, first, select a manageable plot size. "A large space seems exciting but try to be realistic about the time and work commitment. Ask yourself: Will you be able to interest others in the project? How will you maintain the garden over the summertime? Are there facility workers who can help out?" Lynda advises:

- develop a plan for longer-term garden maintenance early on.
- have monthly Garden Days.
- start a Garden Club.
- develop a long list of dedicated, adult volunteers.
- deal with your weeds.

Bright yellow California poppies surround the chain link fence. The garden, set in the midst of miles of asphalt and desert, teaches students a simple and primary lesson of nature. Lynda said: "In this urban space, the garden teaches students that nature is there, that it exists." She said that many of the parents of the students are still learning English and participate in the garden, many with expert farm knowledge. She said a school garden builds inclusive community. In an unexpected oasis, a small garden on the asphalt, Lynda's garden symbolizes an ideal solution.

Middle School, Fontana

Teacher Susan Schellehamer started the garden with students at a middle school in Fontana in 2003. Inspired by her students' enthusiasm and the California state "Garden in Every School" (1995) policy she started the garden with three bags of soil and a few plants.

The garden in Fontana, the former steel belt of Southern California, sits in a pocket of the central courtyard at the school; when classes change, the students working inside and outside pass each other and share garden news. The day I visited there was a group of grade six students who Susan was working with during her prep block. The students all worked with focus and enthusiasm while weeding the area. Their teacher worked in the soil, weeding along with them.

It was obvious that students in the garden truly enjoyed learning or as Susan said, "didn't mind learning." Everything the teacher said sparked their interest and seemed to have a direct connection to the task at hand. *Certain areas in the garden are shadier than others. Keep the worms, but get rid of the maggots. Some plants grow better in the company of other plants.* The students listened intently to Susan's guidance. Unlike the usual routine in middle school classroom learning, the students never challenged Susan to justify the assignment or activity. Students knew what the work was for and worked with purpose, genuinely wanting to do a good job. In this garden, teacher instruction simply allowed them to engage with their project and achieve better results. Most of the teaching was spontaneous, delivered in the form of mini-lessons, all the while the students continued their work under Susan's guidance. The students chatted while they weeded, but they obviously had immense respect for the work, the teacher, and, by extension, the land, itself. Susan said that the students would soon have their performance standard tests and that during such times the garden releases tension and stress.

Susan's indoor classroom is filled with plants, vines, garden supplies and intriguing, interdisciplinary learning stations; the site is an extension of the garden and a place where learning takes place as a process. Like the garden, the classroom is organized around students and set up for spontaneity. I asked Susan, a veteran teacher, what lessons a garden might hold for new or student teachers. She responded: "Flexibility. Also, a garden teaches you to ask for help from others…a garden helps teachers build a supportive learning environment, and that supporting the space, in turn, helps you become a better teacher." Susan said that a garden teaches teachers to let go of "old ways" that might no longer work because a garden teaches "what matters." She said that a garden teaches teachers to get down to basics, to "let what is important rise to the surface."

As the desert wind around us began to pick up the desert dust, Susan's students intently, carefully and quietly weeded the garden. Here, the work of the garden itself, the school, the surrounding community, the work inside the classroom and Susan's commitment to students and to the natural environment all co-existed as part of a positive, relaxed model for learning. The garden represented how a school garden, even in a small scale, prompted slower, authentic learning processes. In an urban context, the garden brought student focus and a sense of care and community.

Susan said: "a garden teaches people how to be human." She never raised her voice to her students, and it was obvious how much the students respected her. I asked the students if they liked the garden. "Yessssss," they all answered in unison. "What do you like about the garden?" I asked. Some were more interested in the food they planted and ate in the garden, others liked the flowers. Every question brought a positive response and enthusiastic nodding.

Susan's classroom has pictures on the walls of students working in the garden the year it was built. She said that former students keep in touch with her, visit and still ask about the site. Between classes, all of the students were drawn to the garden and chatted curiously with Susan and the student workers. Small pots of new plants were waiting to go in. As a focus for community and stewardship in the desert Susan's teaching philosophy in the garden was illustrated: "in a garden, everything matters."

Middle School, North Redlands

Gretchen Noble is a Special Needs teacher at a middle school in North Redlands, California. School garden programmes aimed specifically at urban and suburban youth with special needs is rare and research practically non-existent. Gretchen, a senior teacher with fifteen years' experience, promotes equality as the basis for learning. Such a philosophy easily translates to her work in the garden.

The greenhouse garden (for growing vegetables) began a number of years ago on the site of an existing program. The school's other, smaller gardens, include themed butterfly gardens and native plants. Gretchen explains that vandalism is not a problem because "the students have grown the produce and see the relationship between themselves and their surrounding space. The students monitor each other. They'll say: 'Don't touch that!' It is a 4th year program, so the 4th years teach the new ones....how to shovel, weed and rake."

At this school garden work is shared, with all of the teachers involved. Gardens are in fact tucked into every corner of the school as living, spontaneous proof of the teachers' dedication to the space. The school gardens are set to expand, with a science teacher using the gardens to expand a native plant area, which includes plans to grow plants from the different regions and climates in California. The smaller gardens in the inner courtyard are part of ongoing projects by various teachers, and there is a sense that all of the spaces are in the midst of an experimental, creative process.

The students here, as in many other low-income areas, use the garden produce themselves. Gretchen's focus in the garden with special needs students rests on a combination of health and education. Her focus is to teach students about the origins of their food: "It's not just all about *the store*." Says Gretchen. "There is so much illness amongst children. Teach our kids to eat healthy! Teach them about food origin. They see *cause and affect* in the garden. They learn: "I am eating this same pea that grows from that plant. They see how plants relate to soil, and relate to air and sun." She adds: "The special needs students become very involved in the garden. The garden gives them a sense of community, health, and a sense that they are connected to the earth."

Gretchen admits she has a bit more flexibility because she teaches Special Needs students. Still, she relates all of their lessons to curriculum and feels that the garden enhances the curriculum "beyond words on paper." Gretchen explains her relational teaching approach: "I teach and they begin to take over the process ....they begin to learn..... they realize that plants grow from a stem. They see the weeds at the side of the highway and when they return to the garden they go 'Oh yeah. I saw that plant beside the highway.'" Gretchen's methods for teaching Special Needs students in the garden also reveals an innovative pedagogy and conceptualization that might be interpreted as:

- giving up total teacher control ("I teach and they begin to take over the process")
- allowing students involvement ("they begin to learn")
- giving students room to learn ("they realize that plants grow from a stem")
- letting students learn different sequences, processes, and parts to learning
- connecting students to their local environment ("They see the weeds at the side of the highway and in the garden and they go 'oh yeah. I saw that plant beside the highway.'")

On the question of whether the best teaching and learning in a garden is either pre-planned or spontaneous Gretchen replied:

> Spontaneous. The learning starts from geological knowledge or textbooks of plants—then it works its way outwards. It's all experience and spontaneity. We are responding to what's there. To what's been happening. They know how the routine in the garden goes from there, and takes off.

Gretchen described the spontaneous process of teaching itself as a central technique to garden-based learning. As she stated, the plant knowledge exists, but it is the "moving outwards" that provides the spark that moves a classroom from teaching to learning. When her students arrive at the garden she tells how they simply respond "to what's there" and "to what's been happening." Students' spontaneous responses to local place, time and process is, as Gretchen describes, where her teaching "takes off."

On whether the value of urban school gardens could ever be proven in qualitative or quantitative research terms, Gretchen said:

> The value is proven when the kids go home and can talk about what they did in their garden with their families. Kids and tarmac: does nothing for them! In the garden....these student see themselves as part of nature. And, they are.

Kimbark Elementary School, San Bernardino California

> Dave Goodward's fifth-grade science class lined up across a sunny ball field in San Bernardino earlier this week, counting—and occasionally chasing—orange-speckled butterflies. "I got one! I got one!" hollered one of his students from Kimbark Elementary School, adding another catch to his class' unofficial butterfly census. (After the Rains, Tiny Rainbows. *Los Angeles Times*. 2005)

Kimbark Elementary School, an environmental magnet school, sits near the Cajon Pass between the steep Sierra Madre Mountain Range and the city of San Bernardino. Dave Goodward is a Science resource teacher at the school who conducts and coordinates classes in the garden. Dave is welcoming and eager to share the work of his students. Through observation of his strong rapport with students across the school, and through discussion with other teachers, it is obvious that Dave has a deep sense of commitment to environmental learning. The students clearly love learning with Dave in the garden and it was obvious that the students' care and respect for the garden began with Dave modeling care and respect in his teaching. Observation also revealed how learning in this garden included and moved beyond standardized outcomes and even beyond the notion of a garden as a teaching tool. In observing the class and the students, paying close attention to the interactive process of learning itself, it became obvious that a teacher who loves and respect the land will pass on those traits to their students, as part of an organic learning system. Here, the word "sustainability" is actually lived through the knowledge, care, methods, curriculum, and respect the teachers constantly demonstrated.

Dave Goodward's classroom itself is like a small natural history museum of Southern California. Stuffed taxidermy (deer, elk, owls) peer down from the walls. The counters are a jumble of sinks, skunks, science lab paraphernalia, dioramas of snake habitat and other desert wildlife. On the first day I visited, Grade 5 students were seated around "Mr. Goodward," enthralled in a hands-on Chemistry lesson involving beakers, matches and small smoke explosions. When Dave asked questions to review student knowledge of different compounds, the students' eagerly raised their hands. Every student in the class was fully engaged. "He's really nice," one girl told me quietly. "He's really cool," one boy said. Observing his class, even before we arrived at the garden, was a first-hand lesson in the best of experiential teaching. Teachers and students were engaged; they cared about the process of learning and the environment in which that learning took place.

On the way to the garden Dave walked at a brisk pace, stopping by several classrooms for a word or two about projects that included: an experiment on the counter (tadpoles), an upcoming field trip, a follow-up to a garden class. As a garden teacher, Dave is connected to all of the students and teachers at the school and is able to create a positive learning community. We stopped in to pick up another class of students (Grade 4) where the teacher said, "Dave Goodward really is the best. He is amazing. The kids love him!" She explains how he prepares and cooks produce from the garden with the students, how he has a way of working with the students "that works." On the way to the garden we also meet Janeen Hobbs, Dave's classroom support person. Like Dave, Janeen is knowledgeable, dedicated and supportive of the student work in the garden. Near the California native plant garden, Dave calls out to the students: "Today we are going to learn how to screen compost! The students excitedly whisper: "Remember how we made compost last year?"

At Kimbark smaller and larger gardens are interspersed throughout the campus. A vegetable garden with raised beds sits on the hilltop, native plants grow down the hillside. Next to the butterfly garden, inside a chain-link fence,

FIGURE 4. Student plants tomatoes, Kimbark Elementary, San Bernardino.

Dave stops to teach the class about compost. The students spontaneously respond without being asked. When Dave says, "It's organic." The students volunteer a definition and expand the lesson themselves based on prior knowledge: "The meaning of organic is ....natural things." "Ah, yes," says Dave. "Like when you buy organic food." "Yeah," say the kids, with him. "Let's start back at the beginning," says Dave. "Organic matter is dead plants or animals." "People?" one student asks. "Yes," says Dave. "We would be included in organic matter." "Ooooooh!" say the students, impressed.

Beside the butterfly garden, he teaches the chemical definition of organic matter:

Dave:	Remember back in the classroom?
Students:	Oh! I know! I know! Carbon!
Dave:	Where did that carbon come from?
Students:	Dead plants!
Dave:	And what happens?
Students:	They gets smushed!
Dave:	Yes, buried! Smushed and buried. Smushed and buried. More die. More die. More die. More smushed!
Students:	Laughter.
Dave:	Billions of years go by....and buried matter, smush, smush, smush....turns into coal. It's a little more complicated than that, but that's basically it. Coal is organic matter. How do we give a living plant some organic matter? Wait a minute. Let's go out to the garden. It'll make more sense out there.

Much of the learning in the garden seems to arise from the community of the garden and the work itself. The teacher's role here is to share knowledge that is meaningfully related to the work of the day: he tells the students that the soil is sandy and requires compost for fortification, and thus compost is one activity for the garden. Another task of the teacher is to organize the students in a way that, again, makes sense in the context of the work: Dave tells half the students about sifting the compost through giant metal sifters while the other half works with Janeen at the garden beds. She plants tomatoes with the students, and from time to time gives them instruction: "24 inches apart."

FIGURE 5. Teacher Dave Goodward and students at Kimbark Elementary Garden, San Bernardino.

Janeen says that, in the garden, "the students really teach each other." The students begin weeding and raking the soil, while others measure soil with measuring tape; because all students have shared in every task, the students are always able to help each other, based on their own knowledge and experience. Much of the work in the garden is about negotiating, division of labor, and the practical means for carrying out the work. The students make garden decisions as equal partners, and I observe a great deal of discussion between and amongst students and teachers before work even begins. Once the students understand their task and they are involved in the work, they themselves become the involved experts and take on a teacher's role, pointing out changes in insects, rocks, soil, wind, sun, and shade. In the garden authority is shared, teachers learn, and community is created. Here, the students understand the true value of team work because they all share in the group's vision and know the vision will be carried out in a way that will benefit both themselves and the group. In this way, beyond the classroom and the textbook, group work has direct meaning and benefit to students. Observing the students at Kimbark, everything in the garden seems exciting, especially the aphids who one student describes as being "at war" with the lady bugs on a "battlefield" of fava beans. Here, students are focused and involved. They measure, weed, rake, talk, make metaphors and tell stories to one another. They learn in a variety of unplanned ways.

"Look out!" calls out one student, running past. "The aphids are having a feast!" Other students spontaneously launch into various songs ("Super aphids! Super aphids!") while interacting joyfully and effortlessly with each other and their teachers. "Can we eat the fava beans?" One student asks Dave. Dave good-naturedly picks up a shovel: "Not until we do our work first."

Janeen at the Kimbark Garden: Organic Matters

Janeen describes how the garden is a symbol of learning to urban students in San Bernardino. She explains that one of the true values of gardens is that they teach students about the "the very stages of life." Gardens introduce a way of learning in school that connects "the compost pile to seeds to collecting seeds to planting to harvest." Janeen describes how the garden teaches students how to grow plants and vegetables using organic methods while also introducing an organic way of learning: "We start with the compost; feed it to the beds; plant the seeds; watch the seeds grow." In this way, the students learn about "plants, environment, and life" as the garden teaches the processes that allows students to sustain the soil, themselves and their learning. While the work involves both technical knowledge and manual labour, engaging in the process itself is also a valuable outcome.

Janeen speaks about the garden as a place that moves learning forward, above, and even in some ways beyond the curriculum:

> You see them in the classroom with the textbook where their learning starts out. Then they come out here and it makes sense. Here, they actually put what they are learning inside to good use. Look at them!

While Janeen talks about the garden the students weed the entire area, measuring the spaces between the tomato plants and gently placing the plants in the ground. There is space and work for everyone; some chase aphids, some sing about them. Others quietly measure the beds. Some work closely with Dave sifting compost, discussing organic matter or the characteristics of different plants. All of the discussion from the classroom is applied and experienced within the context of the soil and the learning community.

A garden is a unique space for learning and creating community, for sharing and sustaining the earth, and the question of whether or not learning in the garden is "magical" does not seem out of place here. Janeen responds to the notion that a school garden changes learning through the rhythms of nature itself:

> I *know* it is magical. We have a wildlife butterfly garden which feeds the "Three Bs": butterflies, birds and bees. We have all kinds of plants over there. When you go you out to that garden it's very whimsical, you see the birds and butterflies coming around. It *is* magical....and the students do find a sense of peace out here.

We discuss the paradox of teaching and learning in a school garden in the midst of San Bernardino. A large portion of the students (300+) are brought to the school every day from down the mountain, from the heart of the city. Most students live in apartments in the highly densified urban area. According to Janeen, the garden creates a new and positive world that separates students from the negative aspects of inner-city life. The garden, she says, "keeps them very grounded," and she believes Kimbark is unique in that it provides a safe sanctuary for learning. She describes the school as "an oasis" tucked away from the daily stress of life in San Bernardino. In this way the school provides a place for immersing students in the beauty of the natural Southern Californian environment that students would not otherwise have an opportunity to experience.

Because of the garden, students at the school learn in ways that are different from other inner-city schools in the area. The school garden provides an important example for learning in nature that is very large in the minds of the children. The school garden, the school's focus on environment, gives the students a local experience that is different, positive, and meaningful. Because of their experiences working in the natural environment in the Kimbark gardens, students do not feel they have to fit into the negative aspects of inner-city culture. As Janeen says:

> There's nowhere like *here*. It's a little more worldly when you get down to the middle schools. More challenges. Up here, the kids are happy. It's like the country.…at least, in their minds it is. The kids say, *I go to Kimbark*. And it's like, What's Kimbark? But they don't care, they *know* it's cool.

Janeen poignantly describes how the garden provides a foundation that empowers urban students, strengthens them and brings them a gift of depth and wisdom that will move with the students in the world beyond school:

> I think that if I give them a seed, a strong foundation, they can carry that with them. These kids needs that. In the outside world, they are going to need that. And they will see the difference. I've seen a lot of kids graduate from here, and come back to visit and they tell me: *Hey those kids at that school over there, they argue about the stupidest things!* Later, they see what this school did for them.

On the subject of classroom management, Janeen says that the method is based on a model of prevention that emerges from the values and processes of organic farming itself. Before they take the students out to the garden the students are taught to value and respect the site; discussions are held around themes of respect for every element in the garden.

> We call the grass *mother nature's hair*. So they don't tear up the grass because they're pulling *mother nature's hair* out! We teach them to not step on the beds, or the veggies, we give them a gentle reminder, "This is your garden—please respect it."

In the garden students are involved in every aspect of learning that mirrors the cycles of nature. And once the process comes full circle and students experience the benefits of their work (vegetables and plants, a beautiful garden, a different world beyond the classroom, a community) their work and planning makes sense, in context. Even the respect the students showed for the grass, beans, beds and garden tools has meaning.

> They learn how to tend, harvest and weed the garden. They never say 'Why are we doing this?' because they see the benefits. They pull up the seed that grew right out of the ground, wash it, and eat it right there. It's a reward.

A student arrives, and proudly hands me a fava bean: "Look," he says, "we grew this."

While Janeen and I discuss the garden, Dave is advising, observing and learning alongside the students. The students are completely immersed in this small plateau above the inner city, with a little breeze blowing in the nearby Eucalyptus trees. The garden is filled with laughter, work and socializing that is within the context of learning and life; out here, the students seem happy, connected and strong. Even the most reluctant students seem to brighten up when opening the fava beans. It is obvious that Dave and Janeen see the value of the garden through the eyes of their students; they make teaching seem easy, and learning even easier. A student picks up a big lump of organic vegetation and shows it to Dave. "Is this compost?" Dave responds casually: "Yeah, and it's on it's way to being soil." The student, walks to the compost bin, drops it in. "Then let's keep it organic," he says.

The Kimbark Garden: Pedagogy and Practice

> The garden is coming along nicely, except the wily gophers have eaten most of the first crop of green bean plants. Time to retrofit with bed screening. I just haven't had the time to do the work. We planted peanuts last week, something new for us from last year. Tomato plants are too big for their britches; pepper plants starting to flower. I just ordered tepary beans from Arizona, something I have wanted to try for years. (Email from Dave at Kimbark)

Between classes, I interviewed Dave Goodward in his Science room, beside a row of computers, below a stuffed taxidermy of a Great Elk. Dave has worked as a garden teacher at Kimbark for eleven years, and as a classroom teacher at the school for twenty-three years. During the interview we discussed the garden, its uses, and its benefits.

Dave and his predecessor, teacher Bonnie Trusler, explained how the Kimbark School Garden, now entering its fourth decade, has involved the shared work and vision of several teachers and administrators, over time. In the late 1970s, former Kimbark Elementary Principal Zoneth Overbey established

Kimbark as an environmental magnet school. With the help of Roy Kniffen, a volunteer from the San Bernardino County Museum, various features were added to the playground, including: a pond, an amphitheater, a weather station, a garden and lath house.

In the early 1980s Dr. Darlene Stoner (California State University, San Bernardino) collaborated with Mr. Overbey to develop an educational program where the garden and outdoor features would become "as much of a part of the instructional materials as textbooks, pencils, and paper." Their efforts resulted in a comprehensive environmental education program where students acquired scientific knowledge about the outdoors in an experiential learning environment. Several of the innovations described in the history of the Kimbark garden, in fact, sound similar to new innovations in school gardens today. As Dave describes: "Zon set the tone."

In the mid-1980s, Jim O'Neill, Kimbark's first environmental resource teacher, joined the staff. Jim brought "high-tech" to the "high-touch" program. Computers, software, educational videos, laserdiscs, and a closed circuit video microscope system were added for classroom use. Teachers incorporated computer simulations in the classroom curriculum that enabled students to interact with programs dealing with topics as varied as resource management, urban planning, and wildlife ecology. For example, students monitored the U.S. Forest service weather data via telecommunications and Kimbark subscribed to ECONET, a program that put Kimbark students in touch with others around the world to compare data.

Kimbark has been the recipient of many awards over the years. It was selected as one of the best environmental programs in California and one of ten exemplary programs in the United States by the National Science Teachers Association. It received the Golden Bell Award, the California Distinguished School award, and most recently a Conservation Champion Award presented by Senator Barbara Boxer. The award was in recognition for the students' work in restoring their hillside after fires and mudslides. The garden at Kimbark is clearly a source of pride for teachers, students, and current principal Karen Hoag.

At Kimbark, the main vegetable garden sits on the edge of the campus, an area with high raised wooden beds. Zon expanded the gardens from the original garden to include:

- *The Native American Garden.* Started in 1993 with a grant from California State University, San Bernardino, and supported by the San Manuel, Mission Native American Band.

- *The Ecology Garden*. This area contains native chaparral and sage scrub plants, growing on the slopes next to the school. This garden was started in 1994, to teach students about native plant communities, including the importance of drought tolerant plants to help conserve water. Dave expanded the plantings in this area, and he uses the area extensively, particularly with Grade 4 students, to study ecosystems. Dave explains that their science textbook contains examples of desert, coral reef and rain forest, and the Ecology Garden, filled with native plants, helps Dave teach the students about their local ecosystem as well. (Dave notes that the students especially have fun with the spittlebugs on the California Sagebrush).
- *Wildlife Garden*. This enclosed space is certified by the National Wildlife Federation and includes a bird bath, hummingbird and butterfly plants, and wild grapes for birds. It is fenced to allow undisturbed nesting. Dave says a pair of bluebirds nested in the space last year, and towhees the year before.

The gardens are used by all of the classes at the school with Dave, as resource teacher, in the lead. Dave described a unique project at the school where he collaborated with other educators to set up garden-based-learning field trips for students across the inner city, on a regular basis. The summer program ran from 2003 to 2005. The unique garden program combined ecology and cultural awareness with interdisciplinary activities around local farming. Dave explains how the project also included a number of collaborators from local educators and community:

> Most trips included a morning discussion about the work of Cesar Chavez, followed by hands-on garden work, Readers Theatre, and activities designed to increase their awareness of the difficult life of farm workers. Dr. Lynn Diaz-Rico at Cal State San Bernardino was instrumental in getting the funding, and helped write the curriculum along with Jackie Lacey and me. The afternoon was devoted to southern California ecology utilizing our native plantings; other sessions focused on the local Serrano Native Americans and their traditional food and fiber.

The project came to a close with an end in funding and a new school district single-track schedule, with shorter school breaks interspersed throughout the school year. The changes in organization meant that a pool of teachers from the school would no longer be able to teach the field trips during their extended summer vacation. The project, with a relatively small budget and teacher-donated time, effectively served up to 60 students from the inner-city.

Dave explains that Kimbark is unique because the local students have gardens at home while many other students live in apartments closer to the city centre. Overall, the Kimbark garden helps inner city students understand the source of their food. Dave explains how such understanding is central to efforts to support sustainable systems, and urban students themselves, in the midst of a technology-based culture:

> The more technologically savvy society becomes....the more society is removed from the source of natural sustenance. These kids need to know what keeps them alive. They need a positive connection to earth. And, to themselves.

Over the years, Dave has observed several ways in which the garden helps students develop a sense of "co-operation, patience, problem solving, critical thinking and other attributes that makes a better person." As an urban educator with decades of experience Dave believes that a garden gives students the initial valuable social skills that will ultimately help them better use academic knowledge and training. As Dave puts it: "A person who knows how to think and problem solve can use all of their skills. Part of coming from a disadvantaged background is..... kids need all the skills they can muster. They're not necessarily going to get a job out this garden, directly."

Effective garden-based learning requires time to design and incorporate a well-balanced curriculum. So far, Dave finds the garden useful in teaching Science and Math, where students focus on design and space. However, in an interdisciplinary curriculum, Dave says he could easily see the incorporation of Language Arts, writing, journaling and fiction. While most worthwhile learning outcomes emerge gradually, and perhaps sometimes even invisibly, Dave says the garden also gives students immediate, practical skills: "The garden helps students with testing in many ways—it helps their attention to detail. It gives them a sense of idealism."

If a garden inspires hope and organizational skills in a medley of social and cognitive outcomes, then effective teaching in the garden requires a sense of openness, spontaneity and the ability to learn alongside students in the garden. As Dave (like several other teachers in this study) stated: "A lot of teachers would go crazy in a garden!"

For urban students a garden is also a draw for attending school in the first place:

> If there's a school garden, a lot of kids will want to go school. Especially in a lower socio-economic neighborhood. So many problems get dumped on them. They need a positive place to come to. The setting....is a positive, in itself.

During the interview I ask Dave to comment on what he feels is the most urgent need for learners in the inner city and how a garden might address that issue in the context of curriculum and schooling. Before Dave could reply, a student calls out: "Can you make that good food again, Mr. Goodward?"

FIGURE 6. Students in garden, Kimbark Elementary Garden, San Bernardino

Student Interviews

When the work is finished for the day I ask the Grade Five students about their garden. Their candid responses describe the value and magic of learning in an urban garden in the Southern California desert.

When I ask what they value in their garden they all reply, in various ways, that they love their garden. The reasons vary. They love the garden because it is: "environmental"; "better than what we do inside" and "it helps the earth." The students describe the value of their garden in both tangible, practical terms and in ways that are less easily definable. They discuss the physical labour, the food they grow, the relationships between plants and soil, and the variety of life found in the garden:

> *We like the garden because we get to grow lots of little edible plants.*
>
> *The Garden is better than classroom. You get rewarded with tasty plants like mustard greens.*
>
> *There's no paper and pencil. You actually get to see it.*
>
> *Better to get your hands dirty.*

I ask the students about their biggest surprise in the garden. Their responses relate to variety and use of plants, and also to their sense of awe and gratitude:

> *Edible plants.*
>
> *All the life.*
>
> *Getting to eat.*
>
> *Finding surprises.*

One student appreciated the garden for providing a balanced, ideal learning program:

> *All kids should learn outside. Spend the morning in the classroom. The afternoon outside. Perfecto.*

Students were also interviewed in groups and responded to a series of open-ended questions about the garden.

What do you like about the garden?

Grade Three students focused on practical knowledge, and defined the garden in terms of their own experience and practice. Their responses show an understanding of cause and effect, where their work of planting vegetables brings a goal and a purpose, within a larger process:

I like to learn about fruit, soil, and roots, what's in plants, what seasons to plant veggies.

In the garden I learn what kind of plants there are, different reasons for different plants.

Some of them we can eat—we plant seeds.

Students also revealed knowledge and awareness of variety, comparison and classification within the garden's system:

There are the plants that like the cold—the ones who like the warm.

We learn what plants to eat and what plants that are not plants and we don't eat.

Don't pick unripe strawberries, don't step on the fruit.

Birds and earthworms. Killdeer and gophers. About hollyhocks and killdeer. Vegetables have seasons.

Some plants are healthy, some fatal.

I learned what kind of plants there are. We learned about nut grass. Hot air and cold water.

Omega 3 is healthy for you. Hollyhocks grow bigger than you.

When asked about what they like best about their garden, the Grade Four/Five students responses related to the ways that they benefit from the tangible elements of the garden (i.e., eating vegetables) while also revealing an understanding of seasonal cycles as related to the learning process. The students express their experiential understanding of the site and use words that relate to the meaning (and joy) of hard work, of waiting, of learning, and of protecting the natural environment:

I love learning about vegetables.

I like eating fruit, planting stuff, and waiting a week or two.

I like learning how to pull weeds from the ground.

Cleaning the beds of the weeds and bad bugs.

I like to eat the vegetables from the garden. We take them home and make soup. I learned

> *vegetables need protecting from the animals.*
>
> *Just how much plants help us with food.*
>
> *I like to learn about the sun and water, keep it cool and growing.*
>
> *I like to eat veggies. Then help plant them.*
>
> *Sometimes we eat carrots. We plant corn and cook it.*
>
> *Some animals help the garden.*
>
> *We mix the dirt. I like it.*

For the students, the concepts of learning, understanding and working in the garden are all deeply connected to life and process. The students show an understanding of how their learning in the garden is juxtaposed with their own hard work. And, they like the garden because they understand and respect it:

> *There are bugs that decompose plants.*
>
> *Compost helps plants grow, you use it as soil.*
>
> *Take care of it. Pull the carrots from the ground.*
>
> *The vegetables are really cool. Learn about it.*
>
> *I learned how to compost old weeds. How to take care of vegetables.*
>
> *I can make my own compost at my own house.*
>
> *We learn how to plant corn, mushroom, strawberries, sweet potatoes. That's really cool.*
>
> *I learned how to identify plants. I learned that the animals eat the food. I learned that food is nutritious.*
>
> *I learned how to eat carrots from the garden. Also, planting, mixing dirt, shoveling.*
>
> *Don't kill on purpose. (And don't eat just any weeds).*

Describe the difference between learning inside and outside:

Here, Grade Three associated *what* they learn (i.e., math, reading) with where, and how, they learn. The garden is also associated with learning technical how-to skills at the same time the space brings inspiration. In fact, students often interchangeably identify practical knowledge, disciplinary learning outcomes and social and emotional understanding.

Inside you learn math and reading.

Outside in the garden it's nice and cool, we learn more.

Inside the garden you learn how to plant; in the garden it's an inspiring experience.

In Social Studies and Science, you can eat stuff.

A garden makes learning visible, and alive:

Outside, we can see nature.

You get to see the flowers.

If you were learning about plants you would be in the garden.

I will learn about the garden outside. Outside you can see what you want to see.

Grade Four/Five students similarly showed respect for the way that abstract principles learnt inside the classroom connects to tangible learning outside in the garden. They are also aware that their own learning habits improve outside:

I think both inside and outside. Tis cool inside and hot outside.

Outside in the garden. You get dirty, you learn how to do everything....

I work harder outside. Inside, a lot of things on a paper.

You can see bugs and insects. Outside makes learning fun and we concentrate better.

It's better in the garden... fascinating stuff to discover.

Students also expressed how learning in the garden makes visible the relevant life and social processes that help learning and survival:

You learn things from school, inside. You learn how to plant fruit and when to get it out of the dirt, outside.

Outside in the garden we get to talk. We talk a lot.

Outside is better because it's fresh.

Outside we help the flowers breathe.

What Would You Like to Tell the World About Your Garden?

Here, the Grade Three students began naming the plants, animals and elements, the entire complex system that represents their garden:

Killdeer

Gophers

Vegetables

Seasons

Soil

Fruit

Bermuda grass

Hollyhock

Compost

Rollie-pollie

Gophers eat the fruit or vegetables.

Big. It's really big!

Students described their gratitude and respect for their teacher and for all aspects of nature's systems in their learning:

That Mr. Goodward is the best teacher ever.

The garden gives us fruits and veggies, fruits and veggies give us vitamins.

We ate fava beans.

The Grade Four/Five class were keen to share the value of their garden through the processes they learned, and the food they were able to grow and eat:

The vegetables are tasty and sweet. Some have sugary flavor.

I would tell the world: we have awesome gardens.

We have the best garden teachers in the world.

I would tell the world how to compost old weeds instead of throwing them out.

How to water the garden and the vegetables.

You pick carrots that you can eat.

The students shared how learning and planting in the garden, they were able to see how the garden visibly transformed, and how they were part of that transformation:

> *What is special and cool: we start off planting plants, then, how much fun it is.*
>
> *It's cool planting, you see what lives underground.*
>
> *I would also tell the world: keep yourself healthy.*
>
> *I think this is cool to other kids. (Hey, whenever you plant a seed you wonder: how does it grow roots?)*
>
> *I like to find worms.*

The Grade One class also described their own unique awareness of the processes of nature. Even at the earliest stage of learning, students showed how their appreciation for nature translated into care and understanding for ecological systems:

> *I like nature. I like learning in our school. We get to plant. We get to see flowers.*
>
> *I learned to be careful with bugs because they are part of nature.*
>
> *I learned how to plant fava beans. We pull up the weeds because they suck up the water.*
>
> *Earthworms and rolie-polies live in the garden.*

When asked if they liked learning inside or outside, one student replied: "Outside smells good." Similar to other grades, students saw their work in the garden as connected to their learning. They described growing and eating carrots, snap peas and green beans. Many students at this level were also inspired by the beauty of the garden:

> *I like the school garden because it looks like a painting.*
>
> *I like the clouds in the garden.*
>
> *Another student replied: "We love everything."*

Southern California School Gardens

> I have been scrambling to harvest vegetables with the kids, as lots of things matured at the end of vacation. I did a neat lesson on pH and chemical and physical change with some fifth graders. We harvested tomatoes, and made ketchup, testing the pH of all the ingredients, and analyzing whether all the processing steps were physical or chemical changes. Ate it today with potato chips.... (E-mail from Dave at Kimbark)

The student responses to the questions about the garden were heartfelt, candid, and unique. When B.M. Davis (1901) wrote from Southern California: "Nature study has largely to do with the relations of things" it is obvious to see that the teachers and students at Kimbark have taken up a similar spirit of learning through their garden. Students here learn about the abstract conceptions of nature in the traditional classroom then walk outside to experience those connections for themselves. The result for students is enhanced knowledge of how food grows in their local environment and first-hand experience of the work required to apply their new understanding.

In the story of humans and nature in urban settings, the cultural significance of school gardens in the cities of Southern California begins with their struggle into existence. As evidenced in speaking with teachers and students in the area, the gardens bring a reconsideration of elements of the larger ecosystem in relation to its desert location. Gardens in this area return the local landscape from a "paradise" back to practical notions of land as a resource that brings survival.

The garden reminds students of the basic principles of life, beginning with the soil. The urban school gardens in this area promote equality, access, integrity, respect for the earth, and, a sense of deep connection. The students and teachers in the gardens spoke with pride, a sense of place and with generosity; they were happy to share their gardens, their ideas and their fava beans.

A garden in an urban setting helps redefine urban learning itself. Along with other realities about the inner city, a garden also includes a reality of seeds, plants, work and fresh food. Where one reality is crushing and limiting, an urban garden inspires positive change and life itself. In urban settings, a garden provides a way to dilute negative influences in the same way a good teacher might. In such settings, a garden brings hope.

When Kimbark won a "Conservation Champion" (2004) award, Senator Barbara Boxer said:

> I was so pleased to learn about Kimbark Elementary students' dedication to replanting local hillsides and helping victims of the wildfires and mudslides restore their commu-

> nity. Kimbark students have not only helped restore their environment, but have also given the public important tips about planting in fire areas. (Barbara Boxer, California Senator).

Students in Southern California urban gardens now engage in the practical work of re-building hillsides and riverbeds affected by forest fires. They learn about health and well-being, and express joy in discovering and helping their natural environment. The students find themselves in the resilience of the natural world. Just off the freeway, learning is transformed through a deep experience in a small garden, as a symbol of empowerment, and survival. Here, students now learn how to sustain their local land, crops, hillsides and rivers, as they learn to sustain themselves and their own learning.

· 4 ·

NORTHERN CALIFORNIA

Vision and Practice

> For children, most importantly, being in the garden is something magical. You can teach all you want, but being out there, growing and cooking and eating, that's an ecology that touches their heart and will remain important to them throughout their lives. (CAPRA 2005, P. XV)

An urban school garden emerges from the society, culture and history that surrounds it. The social and civic traditions of Northern California are based on longstanding, complex networks that involve shared visions around community. Community and school gardens, both in practice and in concept, have endured here. The urban school garden in this context emerges from hard-fought battles and changes in perspectives toward the environment that have involved educators, activists, business owners, policy makers and politicians, constantly working towards new ways of preserving the land.

Historically, the social climate of Northern California has long depended on challenges to the status quo where both individuals and communities have refused social complacency. An early famous example was Jack London, whose hard-won entrance to university in 1901 was cut short, when "He hoped to find a world of intellectual challenge and bright ideas and vitality. Instead he found complacent professors and little concern about the outside world. He determined that he would be better off continuing to educate himself." (Trinkl 2009) He then traveled the globe, learned diverse farming methods, and went on to invest profits from his book *The Call of the Wild* into land in Glen Ellen, California, where he experimented with various progressive visions around sustainable farming. He created a pig pen on his farm called the "pig palace" where each animal would have its own stall with space to roam and find food and drink. He invented his own farm tools. He dreamt up new forms of irrigation and believed in the healing power of cactus. A sense of dream combined with hard work and practice is characteristic of the non-institutionalized, experimental approaches to land and stewardship here.

In Northern California an activist tradition of conservation is also tied to the natural beauty and vastly diminished natural resources that once characterized the area. Old growth cedar, some 2000 years old, that once spanned four hundred miles on the California coast line from Big Sur to the Oregon border took loggers only one hundred years to remove. Only 4% of that forest now survives, in the hands of state or federal reserves. (Starr 2006)

From mid- to late 20th century, California experienced increased suburbanization and an ideology of privatization became more and more central to urban life. The ideology of suburbia itself increasingly countered certain aspects and innovations central to historical reform and civic life. A reaction to planned suburban safety and isolation was "new urbanism," a term coined by UC Irvine planning professor Mark Baldassare, where planning involved "pedestrian friendly" cities, more public transit, and a move away from isolated mall culture. Along with strong community support for Northern California's natural environment, a reconceptualization of urban space set the scene for a rise in urban school gardens in the later 20th century.

The history of the cedar tree in Northern Californian old growth forests encapsulates the range of both destructive and transformative approaches to land that has taken place in the region. The Pacific Lumber Company owned three-thousand acres in Humboldt County and was an early example of a socially democratic corporation that helped sustain both its workers and the forest. (Starr 2006) In the 1980s, a Texas junk bond trader, with an eye on the old growth forest, bought the company. The notion to harvest the last remaining large stand of old growth went against every idealistic vision in Northern California. Several non-profit environmental agencies went into action, including a group called Earth First! When the state and federal government stepped in to negotiate, the company owner filed a lawsuit, claiming that the government was denying him the right to capitalize on the value of his land. Earth First! organized demonstrations; concerts were planned to raise funds and awareness for the symbolic value of the old growth. In one famous action, in

1997, Julia "Butterfly" Hill climbed a thousand year old redwood and remained for three years, raising awareness for the symbolic value of the tree.

The story helps describe a clash between two prevailing perspectives on land use: the notion of land as commodity to be razed and exploited for the profits of individuals, and, land as place of inherent value, as earth, that requires protection and stewardship. As Starr describes, "Was there not a point, many were beginning to ask, at which society had the right to question the right of an individual…to cut down two-thousand-year-old redwood trees to make patio furniture and pay off junk bonds?" (Starr 2006, 529) Since that time, the California cedar tree has become a Western cultural symbol for balanced land stewardship. As a California historian put it: "A tree, an increasing number of Californians were recognizing, was not just a tree." (Starr 2006, 531)

Northern California mobilized and began to question the ways the natural resource economy was permanently altering the very land by which it prospered. A new approach to the land emerged where business could take part in sustainable land stewardship. Several industrial corporations (including Mission Bay in San Francisco, The Southern Pacific Rail yards in Sacramento) were transformed into sites that promoted environmental stewardship, generated public tax revenue, and exemplified ecologically responsible business. Such bold ideas and actions have also strongly influenced cultural concepts around land in large cities. By the end of the 20th century the notion of "conservation" did not just relate to species or trees but expanded to include "endangered urban neighborhoods" as well. While late 20th-century technology had increasingly become the primary cultural means of social integration, environment issues began to blur distinctions between regions (urban, suburban, exurban and rural) based on a universal need for conservation. (Starr 2006) In the late 1990s, activities like recycling, composting and water conservation thus created a new unifying concept for culture centred on the environment.

In Northern California, the late 20th century resurgent cultures around the re-cultivation of public space also brought the rise of school gardens. The agricultural tradition, the legacy of social progress, the participation of businesses, and the increase in population, provided ideal conditions not just for the implementation of urban school gardens but for the development of a cultural and philosophical framework around public space. In Northern California, such a tradition had roots. In an early (1910) example, UC Berkeley sponsored an elementary school garden at La Conte Elementary through which the pedagogy and practice of the garden was both studied and implemented; the gardens at this site were in fact reinstated in the 1980s. More recently, "The Edible

FIGURE 7. *In the Moment* by Rick Carpenter. 2002. Garage door springs (each) 70" (H) 20" (W) 20" (D). Sculpture Garden, San Francisco Recyling and Disposal.

Schoolyard" in Berkeley provides an exemplary model of urban school garden practice and research.

The Northern Californian urban garden, rooted in activism and reform, is a response to socially progressive action. The significance of the garden here is its cultural role in a larger network of social and pedagogical evolution. In this region, the garden is linked to a network of change that has long supported values around community, environment, and creativity. The value of gardens in this area are not so much related to the well-known success stories of the larger gardens, which represent the typical evolution of a successful social movement; instead, the cultural significance of the sites is the way the smaller gardens survive as a part of, and as a result of, a legacy of a dedicated community network. In this region the pedagogical value of the garden is rooted in both real garden (the space itself, the practical work of engagement in community and social networks) and the imagined, envisioned garden (the hope and imagination that brings these spaces into being)

It would in fact be impossible to study the cultural significance of school gardens in Northern California without considering the role of such networks. In this region, the surrounding community, the places beside and around the site of study, even the blurring of the notion of "place" itself is part of the story. For this area, which includes larger, established models of successful urban gardens, I purposely visited a smaller garden in San Francisco's Visitacion Valley neighborhood. In a small space, the practical and subtle themes of a garden are made visible and the cultural and pedagogical significance emerges. The symbol of the garden in this area reflects the significant role it plays in a widespread network of social progress and community initiative, such as an art-based city recycling program, and stewardship of a public greenway.

Northern California is particularly characterized by social and community visions and a work ethic of making those visions real. In this region vision and practice co-exist for a common purpose. Here, the cultural conception of gardens links to shared, urban, public space as part of a larger perspective on land, learning and life. In Northern California an urban school garden represents the entire community.

Research Innovations

Northern California is a leader in the school garden movement as it encompasses a strong history of both research and practice. Since the mid-20th century the region has been known for progressive social programs around environment, and is notable for linking academic research to practice, with added support from the Department of Education. A brief overview of these models of school garden research shows difference and some similarities, in the cultural, conceptual, ecological and thematic approaches to the urban school garden as it spread and took root on the West Coast.

Northern California, with its proximity to the Salinas Valley, was essential to the food security efforts during WWII (Lawson 2005) The San Francisco Garden Advisory Committee was established in November 1942 as a network of local colleges, women's groups, school boards, newspapers and various interest groups who provided information about community gardening in urban and rural Northern California. In June 1945, the group held a Victory Garden Fair at Golden Gate Park, attended by thousands, to celebrate diversity in the social and natural environment. After WWII, demonstration gardens were built in all the major parks, including Golden Gate and Mclaren Park. By 1973 the city sponsored a community garden program supported by city property taxes. In 1987 a group called the Open Space Citizens Advisory Committee approved Recreation and Park support for The San Francisco League of Urban Gardeners (SLUG) to build, renovate, and maintain San Francisco community gardens. SLUG was the original organizational hub for urban community gardens in San Francisco and for most of the country.

The socio-cultural impact of gardens is a vital element in the research and in ensuring that the gardens survive for future generations. The Evergreen Elementary School in West Sacramento developed a school/community garden that provided small garden plots to families who were non-English-speaking immigrants, primarily from Hmong and Mien cultures. Previous to the gardens, the families rarely participated in school activities; once the project began, a demonstration garden was planted with vegetables familiar to the Hmong and Mien families. Research on this project noted how the gardens "...raised the self-esteem of the children as well as the non-English-speaking parents who were then valued *as teachers*." (Desmond, Grieshop, and Subramaniam 2004, p. 42)

Back in the Bay Area, other related grass roots community projects forged an evolution in urban ecology and research, including Gardens for the Environment, and The Alemany Farms. Among academic research sites, UN researchers call the Life Labs at UC Santa Cruz a major research model for Garden Based Learning. The California Department of Education has also made several innovative contributions in policy, legislation and research goals that directly link to the learning communities around urban gardens. These include the promotion of school gardens in:

1. building bridges between school and community;
2. promoting the transfer of information from one generation to another;
3. developing environmental awareness in students by caring for a living environment;
4. providing opportunities for cultural exchange;
5. building life skills. (Desmond, Grieshop & Subramanian 2004)

In 1995, former California state superintendent of schools Delaine Eastin began a large-scale, statewide program titled "A Garden in Every School." Between 1995 and 2002 school gardens in the state of California increased by 3000 and their numbers continue to grow. Eastin stated that resurgent interest in school composting and farming was "widening the circle of learning from the school garden to local farms…[providing] a context for students to understand…family farming and agriculture as well as the environment they will inherit." (California Department of Education Report, 2002) Eastin's comments describe a progressive movement from the school garden outward *to* local farms whereas, historically, the emphasis was on transferring the knowledge of farming from rural areas to cities. Due to a combination of policy, civic and state support and research, school gardens have a comprehensive tradition of support in California. In 2007, $10.8 million was allotted to California Instructional School Garden Program grants.

Alice Waters, chef and proprietor of Chez Panisse Restaurant in Berkeley is also a prominent figure in the school garden and organic agricultural movement. She is a co-founder of The Edible Schoolyard, a one-acre school garden at Martin Luther King Middle School in Berkeley, where students and teachers grow, prepare and serve food harvested in the garden. The UN describes TES as "…a model in the education of social responsibility, community participation, and sustainable agriculture" (Desmond, Grieshop & Subramaniam 2004 p. 37) The Center for Ecoliteracy (CEL), a public foundation founded in 1995, is closely associated with The Edible Schoolyard (TES) and acts as a major non-profit educational resource for teachers and researchers. One of the leaders at CEL, Fritjof Capra (2005) notes the value of school gardens as a site for developing sustainability:

> One "classroom" that we have found to be especially appropriate for children is the school garden, which reconnects children to the fundamentals of food—indeed, to the fundamentals of life—while integrating and enriching virtually every activity that takes place at a school. (xiv–xv)

Capra supports integration of gardens into school curriculum where students can learn about food cycles while engaging in the planting, harvesting and other practical work of the garden. By incorporating physical work and farming knowledge with curriculum goals, students engage in all levels of a much larger living system. The CEL provides a centre for teaching and research into the deeper, transformative benefits of school gardening, as a practice that connects students to the environment. More specifically, the Centre for Ecoliteracy has identified "fifteen core competencies" around sustainability education that include open-ended goals related to cognition (including knowledge of ecological systems); emotion (empathy and knowledge of social and ecological justice); action (being able to carry out plans); and spirit (including the cultivation of a strong bond with nature that includes support a student's sense of wonder). (Center for Ecological Literacy 2010).

With pedagogical goals rooted in interaction with the natural world, the goals of the CEL and the experience of TES support a practical knowledge of farming as a global concept for learning. A learning model based on application of practical knowledge of land use also potentially changes learning relationships in a school garden that may include but also move well beyond standardized curriculum. The goal of the original school garden at TES is to work efficiently and knowledgably with the earth's shrinking farm land, and also to shift focus in the total learning environment: from teachers to students to the land in a way that unifies the whole learning community.

Two of the founders of TES formulate teaching and research questions around the garden as a shift in values:

> How do we cultivate in children the competencies of heart and mind that they will need to create sustainable communities? How can we design schools as 'apprenticeship communities' that model the practice of living sustainably? (Barlow & Stone 2005, p.1)

The Edible Schoolyard model conceptualized learning as both a tangible outdoor experience and as a site where the land itself influences the shape of teaching and learning. In this way TES directly challenged 20th-century mainstream methods of teaching and learning by emphasizing the pedagogical potential of practical knowledge and connections to ecological systems. Further, such vision and change took place in the midst of a school garden, far from academic think tanks and universities. As researchers at TES note: "Small changes have profound effects."(Capra 2005, p.27)

In the Bay area, the idea of urban gardens has grown from working the land to affecting learning practices to the expansion of community networks to transforming cultural beliefs. Here, a garden is not a garden; and one garden is rooted in many gardens.

The Virtues of César Chávez

- Hope: To believe that seemingly insurmountable obstacles can be overcome;
- Determination: To stay with a course of noble action despite obstacles.
- Courage: To meet danger or advance unpopular opinions without giving way to fear;
- Tolerance: To resolve differences peacefully by understanding the beliefs of others.

("Homage to Four Virtues of César Chávez." 4 Directions/4 Virtues installation, Chávez Memorial, Cesar Chavez State Park, Berkeley, California)

Northern California's environmental awareness grew through diverse networks, and traditions of community, activism, and agriculture. The rise in urban school gardens in the area is a result of a network of individuals and groups dedicated to visions, practices and ideals. The Bay Area is filled with examples of innovative community actions. In one example, the Alemany Farm in South San Francisco, originally planned as a working urban farm, grows beside a busy freeway and a social housing project where community members share garden work and produce. Mayor Gavin Newsom called the site a model urban farm that provides an exemplary vision for other vacant lots in the city.

Other urban innovations unique to San Francisco include the artist-in-residence program at the transfer station/recycling centre (the first of its kind in North America) and the numerous and well-known Latino murals on city buildings. The murals, like the urban gardens, are tangible cultural symbols of community participation and freedom in the urban commons. The murals, often placed beside community gardens, combine vision, practice and a sense of the sacred to transform previously ignored public space.

While the murals are now a part of everyday life in parts of the city, such places are important for their cultural and teaching roles in reclaiming the commons. The symbolic power of the sites is far reaching: "Put simply, free spaces are settings between private lives and large-scale institutions where ordinary citizens can act with dignity, independence, and vision." (Rivera cited in Delgado & Baron 1986, 17). The murals symbolize altars of meaning where space itself is made sacred. As a cultural embodiment of everyday urban life, such symbols represent everyday struggle and form part of the reclamation of public space in order "...to fulfill its primary functions of nutrition and enlightenment."

(Rivera cited in Delgado & Baron 1986, p. 2) Goals of "Nutrition and enlightenment" were rooted in the work and ideals of Cesar Chavez, a Mexican-American migrant farm worker, and founder of the United Farm Workers (UFW).

By extension, murals on the walls of San Francisco have a similar impact as community gardens. Public murals and gardens and other public reclaimed spaces have significant value as local, cultural, ecological and cultural artifacts. As a material symbol of community, the common garden in the city thus represents both a reclaiming of the commons and a re-imagining of a social and pedagogical transformation through organic processes. The patchwork of involvement evident in the murals reflects a social history of care that relates to both the paintings and to other projects that support the commons. When placed in schoolyards, such symbols become part of everyday learning.

In San Francisco, amidst a context of social activism and strong environmental networks, small urban school gardens play a large role in supporting the basis of social and ecological sustainability. Here, the urban garden is a cultural symbol of imagination, history, and creativity. In San Francisco a garden, a mural, and other projects create magical, self-inventing learning spaces that remain part of a community and inspire a sense of hope through imaginative, creative processes.

Cesar Chavez Elementary School

The garden at Cesar Chavez Elementary School provides an ideal symbol of the vision and cultural value of a school garden in the urban core. The school garden grows along the front of the school surrounded by a bright green chain link fence. A mural depicting learners and books, immersed in nature, covers the front wall of the school.

Garden coordinator and Grade Three teacher Russell McCall (students call him "Mr. Russ") discussed the practical and subtle benefits of the garden at Cesar Chavez. As garden coordinator, a position that is part of a district grant, he works with a small committee of interested teachers who consider various ways to engage the teachers in the garden. Every spring the school has a combined Cesar Chavez/Earth Day that usually involves the school gardens. And while the garden began as a community/neighborhood-maintained site, over time, various waves of teachers have taken an active interest. Russ views an inner-city school garden as a site that supports both students and teachers. He says the garden allows practical application of academic and scientific concepts

FIGURE 8. Cesar Chavez Elementary School Garden, San Francisco.

while also reminding students that they are in fact connected to an ecosystem "called Earth." As Russ suggests:

> I've seen with my two classes that with the garden I teach a lot more biology in Science. Last year I kept taking my kids out and we'd sprout seeds in class, plant them, and grow them, go through the whole cycle. The garden makes such a difference. It's really different than growing plants indoors in a cup!

Russ states that the small urban school garden is important for its symbolic and pedagogical value. The garden is an effective way to teach urban students lessons on the value and power of nature itself:

> Even though our garden's just a skinny sort of rectangle—it's still alive. And it's the earth. It's wood chips on the ground....and the earth underneath. I feel like...it's a completely different environment for them. So even though the garden is right next to our school, right next to the sidewalk, the students act really different out there.

The garden itself has an impact on students and Russ observes that the site also connects students to a larger ecological system of seeds, and life, subject to the cycles of the natural elements. Russ describes how a garden brings a balance of experiential learning to urban students and challenges them to think and learn in new ways, in new locations, in tandem with an ecological process:

> They observe natural cycles as they happen. They find the insects that come in or something happened to their plant or the wind blew it over or now it's in a shady spot or....it becomes more like this...deeper understanding of what a plant actually deals with in the wild, just outside. And I feel like they don't get enough "outside time" these days. Especially living in an urban neighborhood. To just be able to run around and not worry is really special.

Russ feels that part of the power of the garden is that it balances academic learning and with experiential learning. For Russ, the garden effectively allows teachers to practice the call for more "hands on" Science, increasingly required in curriculum guidelines. The garden is especially effective in teaching scientific vocabulary within the growth cycles of nature, as Russ explains: "They look in the book....they see the plant cycles in the garden...and now they can understand it." Russ also describes the transformational potential of the garden as a site that invites students to slow down and take notice of the living systems that surround them:

> In the garden I've seen kids who are less inclined to focus on things....slow down. All in all, the students have a pattern of getting all their stimulus needs met when they're outside. Outside there are a hundred different things to see in one place. So they can *really* slow down.

Russ suggest that valuable learning in the garden invites a blend of organization, spontaneity and practical knowledge as part of an organic process that's constantly changing. He believes that teaching begins with an organized structure no matter the setting. From there, spontaneity in the lesson unfolds in ways that places students at the centre of discovery. Russ describes how an organized lesson usually takes off in the garden:

> They see something....and they tell people. And then they find something new. And I'm sending them in a direction and then they might find five different other things they want to focus on. They might start off watering and end up chasing the moths down the garden. Then they check out what plant it lands on. Then we talk about that. Then they catch them.

The previous school year had the students in a year-long discovery of cabbage moths, and their caterpillars. In an example of a spontaneous Science project, Russ was able to incorporate the removal of the harmful species into his lessons:

> The cabbage moths harm the garden so we needed to get rid of them. We put them in a terrarium. Watched them hatch. And then they had this ongoing [moth hatching] centre.

Overall, Russ encourages teachers to begin teaching in the garden using the simplest possible starting point. He invites the teachers to lead students outside:

> What I encourage teachers to do is, at least, just take students outside on a tour. You walk down...and back. Just turn at the end, and come back....Even that little ten-minute exposure every week, they'll enjoy.

As an educator, Russ believes that part of his interest in bringing students to the urban garden relates to his own background where he grew up surrounded by open wild spaces that easily promoted unstructured play and learning. He feels that simply allowing urban students to access nature is a also a potential way to diminish increasingly apparent fears that now permeate many aspects of schooling and life:

> A lot of people are starting to say....we're living in this very unsafe world now....and its affecting the kids. *Fears*! So, any positive exposure to the world is good.

While "positive exposure" to the garden can include a balance of curriculum outcomes taught within an organized structure, Russ also believes that the true value of the site often comes back to the garden itself:

> I think they learn from just walking through a garden. Especially if it's *their* school garden. They see how it changes over time. They just naturally notice things...and learn from it.

Visitacion Valley

El Dorado Elementary is located in the Visitacion Valley neighborhood in South East San Francisco between the Bay and the San Bruno Mountain Range. The school and its garden sit in an area with a rich history of farming, immigration, industry, gardens, long-buried secret tunnels, visions and, ultimately, restoration of community gardens and a waste disposal site that recycles garbage into art. The story of Visitacion Valley itself is one of transformation.

The name Visitacion Valley was brought to the area in 1777. A group of soldiers and Franciscans became lost in heavy fog between San Diego and the Presidio, traveling the King's Highway. In the morning the fog cleared, and they were able to see the beautiful valley and the blue waters of the Bay below. The date was July 2, the Feast Day of the Blessed Mother, and the friars named the area Visitacion Valley. A mass was held, with a large rock as an altar. The rock now sits in a backyard on Delta Street, near Visitacion Avenue and El Dorado Elementary School.

From that time forward, until 1835, the valley was used as a grazing area for cattle, horses, and other animals kept at the Presidio and Mission Dolores. In the 1920s, the area became known as an area for new settlers, farming and gardening. (Cox, Epps, Fishtrom et al. 2005) Down by the water, a number of companies set up industry (Southern Pacific Railroad, a gas plant, a fertilizer company and Schlage Lock Company) in an area known by locals as "the bone yard."

Throughout the early 20th century, Visitation Valley was long associated with the San Francisco City Dump, which until the late 1960s used the shoreline at Brisbane Lagoon to dump garbage from the city. The area was covered daily with rock and soil quarried from the nearby hills. In total, 500 acres of landfill were dumped into the Bay.

Now, the Artist in Residence Program at San Francisco Recyling and Disposal provides studio space to artists who make use of discarded waste materials in their art. The program, established in 1990, has also rescued plants from the waste site, which are replanted in the sculpture garden. Some of the concrete from the old Embarcadero Freeway, damaged in the 1989 earthquake, was used to make concrete walkways in the garden. At the former site of the dump, "garbage" is now viewed as a potential pathway of enlightenment.

On the hillside, workers at the disposal site have selected various discarded items and have made an impromptu sculpture garden. A papier-mâché tiger, missing its eyes, looms in the grass above the toxic waste centre.

FIGURE 9. Impromptu Sculpture Garden at San Francisco Recycling and Disposal.

In Visitacion Valley, further social changes took place as a result of partnerships between government agencies, urban planning and community groups. When HUD (U.S. Department of Urban Development) demolished social housing projects and worked with City Hall and San Francisco State's Urban Institute, the area was revitalized for residents. In 1995, three public plots were transformed into public space by two women: Fran Martin and Anne Seeman, from Friends of Visitacion Valley Parks. They worked to have the land transferred from the San Francisco Public Utilities Commission to the Recreation and Parks Department. Since 1999 the site has been a public space with an Herb Garden, Native Plant Garden and Community Garden, and pathways, art, and free meadows. (San Francisco Neighbourhood Parks Council 2010)

The approach to land in this region is symbolized through the transformation of a garbage dump into a garden and artist's space. A region that once dumped tonnes of waste-fill into the Bay now invites artists to build garden sculptures of dragon flies from discarded metals. Such projects symbolize a collaborative approach to revitalizion where the local neighborhood itself, instead of acting as a backdrop for high-impact land-use decisions, now inspires learning around environment, art and community. One local educator told me that The Visitacion (the vision), the transformation of the landfill, and the rock on Delta Street, are all integral to the story of Visitacion Valley, and to the garden at El Dorado Elementary School.

El Dorado Elementary School

> At the Garden Project, we are thinking small. We didn't have a choice—we didn't have any fancy tools or equipment. We started with what we did have: land, work to be done, and people eager to do it. We didn't know when we started our garden what impact we would have. We only knew we had to begin somewhere. (Catherine Sneed, Director, Agriculture-Training program for Former Offenders, San Francisco)

Maria Acosta, nutrition coordinator at El Dorado Elementary School, uses the garden as a tool for teaching students about nutrition. Her interest in school gardening started soon after she began working at the school in 2005, when she noticed a small unused garden at the back of the school. Maria began with the materials that were readily available, including plants from local nurseries that were "half alive." As Maria describes it, she did not think too much about the garden when she first reclaimed the space: "I just threw some plants in the ground." When the garden started to come to life her interest grew, and soon she was surrounded by plants in the ground and peas growing in barrels. In a way that mirrors the magical stories of Visitacion Valley itself, Maria says that her vision for the space was soon connected to tangible changes: "After I began to envision the space, things started to come my way."

From her work in the garden she began to connect with like-minded people and networks; Maria explains that connecting to city resources around gardening, youth and urban ecology has been an important part of the process.

> Our breakthrough has been this year (2009). We hooked up with The Greenway and raised some money for them. They became interested, Opal and Erin (two garden volunteers from the neighborhood) came on board, and we've written grants and had endless vision for that space.

Maria explains that the garden has changed how she views land in the city and she now sees possibilities for transformation everywhere: "What I do with spaces now....when I see a vacant lot, my mind just runs and I think of all I want to do." The garden also allows her to give shape to her values and goals for teaching and learning in the midst of a difficult urban space. Healing, gratitude and a deep connection to the earth form the basis of her teaching philosophy around the garden. For Maria, it always begins with the soil:

> I have this very large desire to heal the soil. I want to give nutrients back because it gives me so much. And when I see soil that's all used and tired I just want to put my hands in it. Actually....I want to sheet mulch it! I'm all about that soil—if the soil is good, good things will grow from it.

Maria believes that personal responsibility, rooted in a sense of authentic care, is central to stewardship and to encouraging students to take part in the garden. The garden, the students, the school, and the practical work are all connected as the outward visible transformation of a transformed, interior belief system:

> I have to take care of the earth, and that little patch is what I care for. I also see it as an extension of how I feel about that school. It's what we are all trying to do there—build something up, make the basics solid.

Maria describes the affect of the El Dorado garden on the students and how the affects of a small space, a small idea, grow into a space of well-being:

> This Monday we had the kids come through. They were so good in the garden. I really want to do more garden-based nutrition lessons. That's where it starts: good nutrition. It's like our 'life lab' out there. It's moving along. Even though it's such a small place....it has me doing big things.

In the El Dorado garden, action is rooted in awareness, empathy, care, imagination, inspiration, enthusiasm—with profound affects. As Maria says:

> I do not have a garden background, but I have a giant garden desire. I want to do something for my yard, urban plots, even sidewalk patches. It's something I learned because I started to do it.

Like other urban school gardeners, Maria is influenced by her own early relationship to land. For the school garden, she wants to be able to give students an opportunity to have access to soil as a basis for learning, life and community:

> I am from a desert in Southern California.....I don't remember wanting to plant anything. (I was always into digging dirt and sand though!) I started working in Vis Valley about 5 years ago and I fell in love with it. It's my life and I love the kids and staff. To me, that garden really represents our community, how it grows, and how we take care of each other. It's a lesson out there.

Maria says that the most magical experience in her garden started with a barrel of peas:

> When I saw those peas grow. I had two kinds of edible ones, and the colorful sweet-pea flowers. I just was in love! In that barrel. I took a chance on some half- dead snow peas and things started to grow. That motivated me.

For Maria, the chance to work in the garden has opened up a community of care and knowledge while also providing a place for her, as an educator, to share a part of her self in her work.

Another educator central to the El Dorado garden is Opal Essence, a garden volunteer at El Dorado Elementary School. She is also a local Visitacion Valley artist and community member who lives in the neighborhood. Teacher Lori Fetzer, who formerly taught Grade 5 at the school, also teaches in the garden as part of her new role in "school climate support" where she integrates principles of creating a positive social climate with academic work.

The garden itself is at the back of the school, behind a short, blue, wooden picket fence. On the other side of the garden, a painted mural covers a concrete wall above which Mclaren Park rises up a long, steep hill covered in grassland. The garden includes a mix of recycled, found material, recycled fencing, and long benches (with lids for storage) The garden and the school are perched above the Greenway, Highway 101, the airport, the former landfill, the sculpture garden at the transfer station, and below Mclaren Park and the San Bruno Mountains. In the midst of this densely developed area the garden is hidden and tucked away. The site is as peaceful and quiet as the middle of a forest, even as students pass by the garden all day between classes. As they pass, their eyes follow the changes in the vegetable boxes, then move on up the hill to the park.

Opal explains how the garden in fact started from the mural, painted in 1999, when a kindergarten teacher started experimenting and cultivating the area. The current vision was sparked by The Greenway, a project that began in 1995 when three private garden plots near El Dorado Elementary were transformed into public green space. The Greenway now contains an Herb Garden, a Native Plant Garden and Community Garden, pathways, art, and free meadows. Students from the school volunteered at the site and, in return, The Greenway made donations to the garden at El Dorado. Lori explains that the garden was built by "people power," a blend of volunteers, staff and community involvement.

The first priority for the space was to somehow reshape the garden beds beyond a standard square model. They decided on serpentine or snake-shaped beds to accommodate a class of twenty students. Opal and others also developed an irrigation system that would operate for one hour in the middle of the night (when water usage is least expensive) Due to the small size of the space, old wine barrels were used to contain a fig tree, raspberries, grapes and native vines.

A long-term garden plan was developed, decided upon and proceeded slowly. Opal explains that developing a small space carefully, with intention, was always an important part of the plan. In this way, even a focus on one plant has powerful pedagogical value that can nurture an entire K-5 school with valuable garden knowledge:

> We started with fava beans, hoping to put nutrition in the soil. The garden's been growing for 2–3 months and it seems to me that every class knows about fava beans in the whole school!

Lori says the garden connects the school to the location and helps students learn from the green space in and around a local place. With the Greenway and Visitacion Valley neighbourhood below the school, and Mclaren Park above, the school garden thus symbolizes a bridge between the city's cultivated green projects and the natural environment itself. The garden and the path leading to other public spaces have all become part of the learning in the garden. Lori describes how the garden is balanced between these different spaces:

> The school is connected to the park and the greenway. The place between. A connection. A *connecting place*. The kids could walk all the way to this urban area, down this green path.

The small garden, designed with the surrounding community in mind and cultivated with intention and care, creates an organic, uniquely urban, connection. Lori feels that the connection between place and learning first involves taking note of what is already available in the surrounding area:

> We're up against this park. We're trying to make the school an extension of the surrounding community. We are in a large urban area. But we are also located in a place where nature is accessible. If you went down this back door here, you'd find a path and a big park.

In addition to linking students to the surrounding area, the garden also invites students to consider the ground below them. As all of the educators at El Dorado suggested, connecting students to soil is important for urban students with limited access to land and to easily accessible, fresh, local food. Every aspect of the garden was designed in a way that brings students closer to the ground level:

> When we first started, the kids didn't want to get near the ground. Maria wanted to connect the kids to the soil. So we came up with these little seats made out of tree stumps.

They found old discarded benches, painted them blue and used old rubber automobile tires for hinges.

It's obvious that care and thought has gone into the design of every inch of the El Dorado garden. As Opal says: "We're just starting out...and we started with where the kids actually are with the land." While the garden began only one year before, the transformational affects in the students are already visible. Lori says: "The kids already see the garden as a way to be powerful. They can do something and then see the results...they really see that they have some positive power." As "social climate co-coordinator" for the school, the garden helps Lori integrate the social, emotional and academic aspects of learning: "The garden brings cohesion to the different parts of learning; it's all interconnected."

Opal noticed how choice of plants and design may have facilitated student involvement in the garden. As Opal explains it, "as the fava beans grew, student interest grew":

> Because of the new serpentine shape of the beds, the garden became more intriguing to the students. Once they actually got down to the ground...to feel what was growing at the level where the bean is...well, now, they want to help. Now they always ask: 'What are you doing? How can we help?'

One specific incident helped Opal realize the power of the garden:

> There was one little boy helping me a few weeks ago. At first he was very uncomfortable. He was like, 'Ewww, the ants.' Didn't want to be here. The group of kids left and at the end of recess Maria came back with the boy, this time 'in trouble' for some incident. Maria spent time with him working in the garden. Soon he actually wanted to sit and shell the fava beans! Some other boys arrived and sort of mocked him, 'What are you *doing*?' He was like, 'I'm cool. I'm shelling fava beans.' He also admitted that he was disruptive in class earlier that day. It was interesting that the garden was the place where he could calm down, where he 'stepped away.' It was here...that he wanted to participate. To shell fava beans. He could then see his actions clearly. Maria explained to the student that not all behavior should be rewarded with a visit to the garden, but, in this case, it was his way of calming down.

In this way, Opal considers the garden a place that potentially brings healing. She describes the strong school/administrator support from Principal Tai-Sun Schoeman who has helped the garden expand beyond the original vision:

> As you can see, we have a very concrete looking school. But, we have a whole alley way over there, and we're going to bring in soil and compost and put it right on the concrete. We have four recycled wine barrels and boxes. So we are starting to collect

> things that will take the garden down that concrete alley. The principal made the suggestion that we use chalkboard paint so that the kids can decorate the concrete walls, and make it their space.

Like Russ at Cesar Chavez, Opal says the best lessons are simple yet far reaching: "Their sense of curiosity is so strong. They want to know more. They just want to know more and more! It's like you're opening up a door and their curiosity is just endless."

While we talked, the students passed by, roamed, and glimpsed the goings-on in the garden between classes. Opal and Lori agreed that the garden is inspired by the school's surroundings of Visitacion Valley with its vital community tradition that unites people, personal health and public space. The garden educators often bring students to the Agricultural Garden at The Greenway where they pick lettuce, carrots, edible flowers, then return to the school kitchen to cook them "and use them right away" in a salad or soup.

Lori takes the idea of student engagement to an even deeper level and describes the way a natural setting creates a changing dynamic of learning roles that is connected to a larger, organic world:

FIGURE 10. Student in the garden. El Dorado Elementary School, San Francisco.

> I feel like the students can be themselves in the garden. They're not trying to fit into the role of the four walls. It takes a bit more energy to fit into the role of the classroom. It take a lot of energy to say: I'm a student right now! And this *is* a classroom....but it is a classroom with a bit more wiggle room.

Lori explains how the garden, as a different learning space, invites roles and ways of learning that are "situationally appropriate" for that new space. By altering expectations of what is in fact deemed appropriate many students in the garden then find it easier to act "appropriately" and make a positive contribution to a learning community. The changing dynamic of learners in the garden also has an impact on teachers and teaching. In her experience as a garden educator, Lori has noticed that a fairly large number of teachers feel comfortable teaching in the garden. For these educators, the garden simply inspires a continuation of the way they already teach, going "with the flow" of teaching and learning:

> That's the kind of flow they have in the classroom, and they know that's the kind of flow they can have in the garden. Sort of...meeting the learner right where they're at.

Lori believes that a garden opens up the concept of learner to include the teacher as well. In fact, a garden is a place where the students themselves act as change agents in developing a new culture of learning:

> I think a lot of teachers in the garden feel like—they're the learner too. They learn from what the kids bring to the garden; [the students] bring natural curiosity. You know how kids sometimes push the boundaries a little bit....they always try to find another way to think about things.

When the place and the learning dynamic shifts, when teaching and learning moves beyond the usual comfort zones, the students take on new roles that also invite teachers to step outside of their comfort zones. In this way, "the teacher becomes a learner." On whether teaching in a school garden is spontaneous or pre-planned, Lori explains how a certain compromise between both leads to the new ideas.

> There's definitely set ideas about what will happen but there's also room for what will naturally happen. Kids *always* take it further. From the structure, there's a base.

When place itself becomes part of the lesson, learning takes a new direction that involves students in ways that are usually unplanned, and fundamentally basic, yet also central to connecting students to the ground:

> Sometimes the new space....*is* the lesson. I think it's important to realize that. Sometimes that *is* the agenda: to just come out and observe. Or just come out and be mindful. Or just come out and journal. Or just come out and soak up Vitamin D!

Opal adds that the way to "proving" the value of urban school gardens starts with asking the students, and in many cases the answers begin with food:

> Once kids see they can eat things it just goes from there. Their natural curiosity is sparked and doesn't stop. They've been putting things in their mouth since they were little and told to spit it out. Here, they can put things in their mouth and actually eat it. And it also works with their sense of community. When I'm here the kids come by and ask if they can help. Before they knew it was a garden, before they had worked in it and knew the beans were here, they didn't ask to help. Because they *didn't know*. Now that it's "a garden" they say....*Can I help?*

Particularly in urban settings, the key to effective teaching and learning activities lies within the students themselves. Finding the motivation for urban students may not be easy to define because the problems can be complex, the need can be so great and also because such needs are usually hidden from the goals of mainstream teaching and learning. Learning in a school garden thus addresses elemental needs of urban areas. As Lori explains:

> Everyone wants a reward, right? The reward can be pleasure. And the reward can be power. When they work in the garden, they feel both. They're not looking for anything else. I think...in this [urban] community...some of these kids just feel so powerless. So they try to get their power in negative ways. It surrounds them in their neighborhood community. And it can't help but spill over into the school. I mean it's not something so easily turned on and off. It's a coping mechanism for their life. So to give them a tool where they can have positive *power* in addition to pleasure...really enjoy learning...is amazing, really. It all goes back to resiliency.

We gaze at the garden, at the serpentine garden bed, at the hillside above, and the students wander in for another lesson on fava beans.

The Students

In the garden, the Grade Five students walk around, chat, pick and shell beans. The students all know what to do. They speak with excitement for their work, with a sense of urgency. As was the case at other gardens, some spontaneously break out into song as they describe their thoughts and feelings about the space. The student art work and writing in the garden depicts their understanding of the power of food in their lives. The students grow, harvest and cook spaghetti squash, peas, pumpkins, onions tomatoes, potatoes, and carrots. Nutrition lessons focus on the sun's role in the lifecycles of plants.

What do you like best about the garden?

Grade Two students responded to this question by naming various, tangible aspects of the garden:

It's outside.

The water.

The apple tree.

The plants.

Eating things.

The flowers.

Looking at bugs.

The fava beans.

Students understood that rules were needed in the garden in order to respect the processes of nature:

Don't kill the bugs (even spiders).

Fruits have seeds. Fruits come from flowers. (So don't pick the flowers.)

Things grow.

You can eat peas.

For this question, the Grade Five students discussed how "having fun" linked to "helping the environment." Students are also aware of how others have helped create the garden, and how the site nurtures their own health.

They make it so we can be healthy here. (The garden has a lot of Vitamin C.)

The students note that the garden teaches them ways to help the planet, in a way that also helps them.

Recycle! Compost! Beans, we can eat it.

Compost. Live it.

Let me see now. Let me see…now!

Describe the Difference between Learning Inside and Outside:

The Grade Two students described how the garden opened a world beyond the classroom:

I can see things in a garden I cannot see in a classroom.

Then, they broke into song:

Every school should have one.
(Yeah!)
It's a special place to learn.
(Yeah!)
We can grow food there.
(Yeah!)

Grade Five students overwhelmingly chose learning outside, while revealing their understanding between "the planet" and "the plant." When it came to preference between indoor or outdoor learning environments, students here began discussing their preference for the tangible elements of learning.

Outside! Outside!

Why? We need plants and gardens to help the environment, to help the planet, to help the future.

Planet, plant. Planets and plants!

If we didn't have plants the air wouldn't be good. We wouldn't have trees.

Students also described how the garden gave them a unique awareness of their place in nature, on a much larger, planetary scale. Yet, students were very aware of how their learning always related back to the soil, the vegetables, and "the bugs":

Outside! The garden has all the self with nature.

All the plants and all the nature and all the bugs.

I like it. Without it we couldn't have food or vegetables.

The students here reveal an awareness of the value of a garden as a symbol of the planet, and of survival:

Yeah! Without the garden we'll die.

We can grow the food and cook it. Hey! We can grow food right here. Right here. Just eat it right up.

What's that? (Opal: it's a bean) Let's cook it.

What Would You Like to Tell the World about Your Garden?

Students were enthusiastic about the garden, and about the garden educators. They were proud to tell to "the world" that they help the environment at their school. They expressed fear that the garden would not last or would be subject to elements beyond their control. (In this case, fire) Most of all, they just wanted to let the world know that they loved the garden, a theme they repeated throughout the interview.

The garden is very great at El Dorado school. So are the teachers great.

World: we have plants. We have a garden. And we help the environments.

I love this garden, world. And we help the environment for plants. Oh, and I love this garden!

You love the garden? I love it. I love the garden. That's all I want the world to know.

If this garden catches on fire. I'll be so sad. (Yeah) Anything bad happens it'll be so bad. (Yeah)

When I asked if there was anything else they would like to add to their responses, one student replied (with much agreement from his classmates):

World: I love this garden.

El Dorado school is perched in a hot, hilly neighborhood, between mountains, highways, parks, greenways, sacred rocks in backyards and garbage companies that sponsor the transformation of trash into art. Before I leave, two girls stop by to offer Opal a lady bug. Opal places the lady bug on the fava beans. It flies away.

Conclusion

We have put up our sail. And we're not just growing plants—we're growing people. (Sneed 2000)

The story of Northern California school gardens is one of balance between social and ecological systems. In this region the idea of nature in an urban area includes a re-invented social environment where nature is part of life in the city. In this region, environmental stewardship and deep consideration of nature has transformed the city. The urban school garden is a living symbol of that transformation.

FIGURE 11. Students shell fava beans. El Dorado Elementary School, San Francisco.

At El Dorado, Maria described the garden as a "life laboratory," In a small space, a living laboratory is a place for watching the peas grow and for linking small, significant events to larger values and perspectives. An urban school garden, through its seemingly improbably existence in the midst of the inner city, invites a community to nature, to transform itself. The words, knowledge and ideas of students in the inner city are also validated through experience in the garden. An educator's love for the earth (as Maria said, "I have to take care of the earth, and that little patch is what I care for") are represented in the lessons, inviting students to express their love and understanding of the natural world. In turn, students experience a wisdom in natural processes and learn that they are connected to the planet. An urban school garden gives them a place.

What difference does a small garden make? The seed of the garden at El Dorado was a mural, a creative, cultural artifact connected to the city that surrounds it. The school had always been connected to the park above them, and to the Greenway, below. In inner-city San Francisco a school garden is a bridge and a visible example of schooling that is deeply connected to place. Like the mist clearing from the Bay during the first Visitacion, a garden helps make nature visible, where it was previously invisible. As Lori said, the Greenway leads the students to the city: through an open door, and down a beautiful country path.

As for student transformations in the garden: the space itself set the scene for students' needs. The serpentine shape was carefully considered to include student involvement. When students enter the garden they learn independence and realize, in that space, they have room to grow. As so many urban educators know, student engagement does not always take place in the traditional pathways. The garden gives the students a practical and empowering role that also gives teachers a chance to teach through experience. At El Dorado I saw the values of freedom and creativity at work as a garden of educators involved students directly in their actions and work. A small garden, one small change, was the catalyst.

As Russ at Cesar Chavez said, an urban garden may be small "…but it's still alive. And it is still the earth." A small urban school garden represents a way of engaging in environmental education where small actions emerge from community, networks and multiple pathways. In such a setting students learn how to cultivate land and they also learn a new way to learn, inspired by place. In an inner city a garden becomes a pathway to knowledge and to an experience of the natural environment that would be otherwise unavailable to urban students. The community pathways in and around an urban garden invite a new concept for learning as process that, in the end, always leads back to the garden.

> The garden team made sure the irrigation was working for the summer and we came back to a thriving, healthy garden…a section of the school wall was repainted with chalkboard paint. Info about composting and the garden are written on the wall in chalk. Maria is doing an after-school garden…The children plucked kale leaves from the garden, washed them, tore the kale into bite-size pieces and put then into an electric frying pan with oil and chopped garlic. We added tomatoes and mixed it with pre-cooked pasta. We placed our meal on bio-degradable plates with bio-degradable forks and went back into the garden to eat our fresh meal and write an article on the benefits of composting for the local paper "Grapevine Visitacion Valley."
>
> Many of the kids had never eaten kale, but were willing to try it because they prepared it. It was a wonderful lesson in the cycle of healthy soil, plants and happy bellies. Maria prepared for and taught a simple lesson….but the happiness of a full belly says it all. (E-mail from Opal at El Dorado, October 2009)

· 5 ·

THE NORTHWEST

Practice and Spirit

> Cascadia is a land nourished and held together by the sacred salmon cycle.
> (CARR 2008, 128)

Environmental stewardship in the Northwest is based on a historical tradition that blends values around the land with innovative practice. Here, cultural understanding of the land is rooted in bio-regional support for local agriculture, especially through an expansive network of small farms and community gardens. At a school in South Seattle such ideals are embodied in a school garden that has managed to develop both a high-functioning, inspirational learning community and a small economy. Here, urban students are involved in learning, trade, and agrarian ideals, some based on practices in their family's home countries and others adopted from local practices. A brief history of the region's approach to environmental practices and programs will help describe the way that a small urban garden exemplifies the deeper goals of ecological and social justice in action.

In the 1960s and 70s, the urban Northwest was considered the unofficial capital of the environmental movement. The area had been known for its natural resources (including fertile agricultural valleys, dramatic mountains and coastline) as well as for its thoughtfully planned urban areas (electric rail in Seattle; free public transit in Portland). The combination of natural beauty, social/environmental planning and strong civic traditions led the way for a strong, regional environmental movement that lasts today.

In the Pacific Northwest, the Olmsted Brothers particularly contributed to the tradition of conservation ethics through designs of various parks, estates and colleges. In 1903, they created a master design plan for parks in Bellingham, Portland, Spokane, and Walla Walla. The Olmsteds designed the grounds for various institutions, including the University of Oregon and the campus of the Washington State Capitol. (Mendelson 2009) In the early 20th century, the Northwest also pioneered "Woodlawn," an early, urban, multi-purpose school garden in Portland that attracted members from the local community and beyond. In 1914, Alice V. Joyce, a teacher at Woodlawn school wrote about the activities of the students "as well as the industry displayed" in *The Nature Study Review*, a research journal published out of the University of Chicago's Department of Education. The description reveals a well-planned, progressive, innovative place filled with pedagogical experiment:

> Among the special features were "The Variety Gardens" (vegetable and flowers), the "Industrial Garden" where sixty varieties of grains, grasses, lentils, forage, and fibers were planted, "The Exchange Garden" where surplus plants from thinning were transplanted and exchanged, "The Old- Fashioned Flower Garden," "The Twentieth Century Garden," and a "Sand Garden" for the kindergarten visitors. (Joyce 1914, 276)

The organization of this early urban school garden, with a focus on food, flowers, industry, and early 20th century experiential learning, reveals a distinctly modern approach to public land with perspectives shifting toward multi-use, teaching and research. School gardens were already beginning to decline by the early teens; perhaps with a movement in decline, the way to maintain relevancy was to experiment with multiple purposes that would serve multiple audiences. However, the more hidden possibilities in the development the school garden were also gaining recognition. As Joyce describes, hope for the garden rested in the leadership of the school Superintendent, "a recognized leader of industrial progress." The superintendent, in fact, recommended a learning plan that began with each local garden and involved: "the natural development of the garden for each school when parents, teachers, and pupils have awakened to the possibilities of the garden and the actual results to be obtained. (Ibid., 280)

The writing style in the work is representative of the era; however, a unique, combined emphasis on both open pedagogical possibility and structured curriculum began to shape the culture of learning in Northwest school gardens. In this region, as far back as 1914, there was a defiant emphasis on respecting nature's known and unknown qualities. In the literature of the time there was a hint of acceptance for the idea that nature held its own distinct knowledge where "school gardens will not be absorbed into an artificial 'City Beautiful,' but, with their throngs of happy children among birds and flowers, will become 'The Real City Beautiful.'" (Ibid.) The journals included descriptions of the localized learning of urban areas, with lessons developed around community and nature's own growing cycles. Learning outcomes focused on experiential learning and a practice called "concrete teaching." Faith was considered a "real lesson of life":

> The teacher may develop many lessons of faith by encouraging the child while the plants are growing. The lessons of mathematics become real problems through actual measurements and calculations. The real lessons of life may be taught through concrete teaching. (Ibid., 279)

The notion of combining practice, pedagogy, personal resilience and philosophy in a school garden was backed by C.F. Hodge at the University of Oregon in Eugene, who believed that nature study as a unique part of standard curriculum "provided a vitalizing leaven of reality and that this real, active problem-solving is absolutely essential to the best moral and educational life of the child in every school grade." (Hodge 1914, 336) Gardening in the Northwest was thus seen as having a practical purpose in the world, one connected to larger community, where students planted, harvested and sold crops at the local market. (274)

Gardens in the region, particularly Woodlawn in Portland, Oregon, also inspired new ideas around localized curriculum and localized learning (a notion still considered progressive today) In one example, J.A. Churchill (1914), Oregon state superintendent of schools, initiated the idea of focusing school gardens on the study and teachings of native plants:

> The subject matter of such a course of nature-study is….not the lions and elephants of Africa, nor the stars and comets, but first of all the pets and domestic animals belonging to the home, the birds, frogs and toads, gophers, moles, rats and mice, the insects, injurious and beneficial common in the neighborhood, then the common plants, the garden vegetables and flowers, weeds and wild flowers, fruits, nut, ornamental and forest trees, vines and shrubs. (367)

The choice of incorporating local moles, mice and insects into garden learning over ornamental vines was (and is) progressive pedagogy. A variety of teaching mottos from educators and policy makers in the region also describe a garden as a means of supporting ecology-based philosophies of teaching and learning. One education research journal supported new experiential perspectives on nature study, developed by teachers who had worked with students in gardens, first hand:

> Teach the child, not the subject. (Churchill)
>
> Study nature—not books. (Agassi)
>
> In nature-study, telling is not teaching. (Trafton, cited in Churchill 1914, 367)

While the measurement of a garden's "moral purpose" was standard in East Coast research surveys of the time, the Northwest research was unique because it also hinted at a modern-day notion of support for the land for its own sake. A Portland Schools survey cited the need to promote local learning as "practical, concrete courses of nature study" in order to support the following goals:

- to relieve from bookishness;
- to develop powers of reasoning, thinking and observing;
- to discover natural aptitudes of the child and bring [them] into harmony with the laws and forces of nature;
- to enable the schools to meet the charge of general apartness from life and futility so persistently urged against them. (Hodge 1914, 336)

At the time, it was a revolutionary notion that nature itself could provide beneficial learning beyond established curriculum and beyond what the teacher could provide. Even as interest in gardens waned, educators viewed such sites as a way of establishing credibility and relevancy for a school system increasingly under attack for not providing curriculum that seemed relevant to daily life. Nature study connected students to life in a way that combined abstract and experiential learning. The urban garden in the Northwest provided balance between the two worlds.

At a time when individualism was highly prized and students often gardened in small individual plots, the pedagogy of the Portland schools was notable for recognizing both the individual learner (Woodlawn supported student development of self-made learning plans) and the group dynamic. As Joyce described: "The pupils are taught that social adjustment is essential to life, that their education consists not only of thoughts as presented by others. . . ." (Ibid.) Her piece "Valuable Lessons Gained Through Gardening" describes the kind of pedagogy that was valued at the turn of the last century in the Northwest urban school garden. The lessons achieved in the garden, also called "a living laboratory of nature," included:

- The power of invention while studying nature;
- Many lessons of economy by saving seeds, conserving strength;
- Supplying the home with vegetables of the child's production;
- Lessons in the beauties of nature through harmony of colors and simplicity of design;
- Self-respect in working at a self-appointed task;
- Dignity in labor worthwhile;
- Perseverance and faith in replanting of seeds and awaiting their production;
- To be well and happy are the lessons of life.

(Joyce 1914, 280)

The lessons emphasize a holistic blend of outcomes that include creativity, industry, aesthetics, design and respect for self and community. Once again, many of the ideas mirror innovations that are still supported today. The literature also includes the often repeated economic goal of teaching agriculture to urban youth in order to re-populate rural areas. However, another new perspective emerged in the Northwest region: a garden aided a "great spiritual aim" which, unusual for the era, did not necessarily depend on organized religion but on a more informal notion of bringing "the child into sympathy and harmony with Nature." (282) Teachers were also advised to engage in systematic experimental work for the purposes of inspiring in students a sense of the "power" of nature and not "just facts." (Ibid.) The blend of practical purpose, industry, experiential learning and open-ended spirituality and respect was historically unique in garden-based learning.

As superintendent of Oregon schools, J.A. Churchill (1914) also pointed to the progressive notion, unique to the Northwest, that while teacher knowledge was limited, nature held unknown answers:

> The bane and distress of so much of our school life is the foolish and impossible notion that a teacher ought to 'know it all,' ought to be able to answer every question about everything. In nature-study we have a subject that instantly shatters this deadening fallacy and gives us comfortable, vital honesty between teacher and pupil. (368)

In addition to recognizing the limits of teacher knowledge, the school superintendent promoted the idea of "not knowing" and humility as the basis for teaching and learning. In Portland schools at the turn of the last century, close interaction with nature was considered a way of opening knowledge to new directions and pathways of true discovery:

> As long as a teacher is afraid to say, 'I do not know,' they will be afraid to even try to teach nature; because nature is infinite in every direction and man is finite. The first

> question a child may ask—outside the limits of a book lesson—is likely to be one, the answer to which [no one] on earth knows. (Ibid.)

Ideas at the time thus began to gather around a respectful, non-hierarchical connection with nature that could be found in a garden. In the Northwest, well ahead of popular opinion, senior educators uniquely acknowledged the limits of human knowledge in relation to nature and supported such ideas as central to teaching and learning. As Churchill noted: "Nature study will vitalize and humanize and leaven the whole lump of our public school education." (Churchill 1914, 369)

Environmental Middle School

> Developing a sense of awe and wonder about nature—that is, appreciating the poetry of nature—is just as important as comprehending its science. (Williams & Taylor 1999, 83)

The first environmental middle school was launched in Portland in 1995. Research from one of the founders of the garden, Portland educator-researcher Dilafruz Williams, shows the evolution of garden-based pedagogy from the turn of the last century to now in ways that are unique to the Northwest.

In spring 1995 the Portland School Board supported the proposal for an Environmental Middle School (EMS) as a public alternative school in the Portland School District. The proposal was launched in the midst of program cutbacks and teacher layoffs; as a result, the district was inspired to show support for "alternatives" in programming. The EMS was indeed unique for a time when no other schools in the district developed ecology- based programming. The school thus opened in September 1995 "...with the mission of weaving ecology into education to make schooling meaningful for adolescents." (Williams & Taylor 1999, 79) Teachers, parents, students, and community members came together in what Williams calls a "...fluid and loosely knit collection of dreamers and planners . . ." (Ibid.) The project included a Native American elder, whose central involvement brought traditional perspectives on the sacred relationship between land and people.

The EMS project encouraged direct experiences with nature and was intended to give urban students a sense of civic pride that would be "nurtured by connecting youth to the bioregion." (80) The project would deal squarely with student high-risk behavior but also the students' sense of commitment to community; learners were encouraged to take responsibility for the city of Portland, and to integrate local urban issues into the curriculum. The school moved the teachings of the school garden from the "margin to the centre of the educational experience." (Ibid.)

The practices of the project were rooted in high ideals. The EMS is significant not only in the history of urban school gardens but in the story of environmental education. The transformation of the whole learner included knowledge, practice and curriculum, and a deepening environmental awareness. The work at EMS was also part of a Northwest movement that blended vision and practice where the natural world inspired bold dreams of social and ecological justice. The holistic blend of experiential learning and discovery in nature, in allowing students to explore the hidden meanings of the natural world was central to the development of a new environmental school ethic. The school included "field studies and interdisciplinary curricula along with…a holistic understanding of the relationship between self and environment" and aimed to develop environmental learning that inspired awe, wonder, poetry and appreciation of nature. (Ibid.)

Early on, researchers and educators at the school incorporated into curriculum the idea that environmental education was the basis of interdisciplinary learning. The school garden was taught in a context of urban social complexity that included scientific knowledge of the earth *and* holistic interpretations of the environment. The location of the school in Portland, a city of parks, public transportation and progressive urban planning, also helped cultivate and nurture the purposes of the garden. The school made a dedicated effort to realize the goals of Orr, Berry and Snyder "to become a 'dweller'" where "environment…is not simply about 'enjoying' nature; rather it is also about recognizing how we are implicated in the use of resources." (91) Such work clearly influenced research methods: "Student responses, whether in the form of poems, essays or stories clearly indicate their connection to nature and to their own bioregion." (Ibid.)

One of the most radical and innovative aspects to the origins of the EMS was the idea of having students simply sit in the garden, with paper, and record their ideas and feelings. The result was pictures and poems that captured both their feelings about nature and also about local growing conditions, plant knowledge, and the harmful impact of humans on the environment. New to this research was the idea that experiential, firsthand narrative, slow contemplation of nature, and urban student voice all have value in developing local, urban ecological literacy. In the previous century, on the East Coast, many urban school gardens had been seen as ways of helping students enjoy (and train for) a rural ideal; here, on another coast, almost one hundred years later, a participatory, grass-roots model of learning had begun in the garden. In a time of cutbacks, standardization and growing bureaucracy in Education, such research asserted that "concerns....about dwindling resources and loss of community in urban areas...are not separate from environmental concerns....the soils will need to be worked and enriched even as environmental education grounds itself at the centre of education." (Williams & Taylor 1999, 100)

A recent case study of Seattle's community gardens reveals more specific lessons for urban school garden research. (Hou, Johnson & Lawson 2009) Among the many findings of the researchers was the idea that community gardens, gardens in public spaces, allow for multiple expressions of urban sustainability that include the visions and work of planners, policy makers, designers, volunteers, non-profit agencies, and other professional involvement. Community gardens grow through the lessons of many working together on projects that may have simply started with a vacant patch of land, an opportunity, or simply with one person's vision. The authors of the study also note that while there are similarities between community gardens, each site is unique to a specific region, due to a variety of factors that include the particular participants, the urban community context and the site itself. In Seattle, and in Washington State in general, a vast network of community, school and agricultural groups are all committed to supporting the cultivation of fresh, local food. With scarce urban land, and fewer financial resources, even with the high level of civic, community and professional support, the success of such gardens still very much depends on the capacity and dedication of the grass roots.

The Northwest brings a unique blend of practice and ideals in a way that applies deeper ideas on environmental philosophy. Such changes relate to the development of a new environmental consciousness that truly attempts to balance the requirements of humans with the urgent needs of the earth that sustains life. As one Portland environmental educator comments:

> Among the most daunting challenges of our era is the task of bringing about the transformation of consciousness that will be required if we are to move away from a culture predicated on consumption and the values of the market toward one that strives to balance human activities with the requirements of the natural world. (Smith 1999, 207)

In this region, non-formal, community-based education is supported as a means of moving adults toward a cultural transformation centered on protecting the environment. Such an idea nurtures a blend of philosophy and practice within both educational institutions and the surrounding community. Such a shift in environmental consciousness is in some ways similar to Frieze's emancipatory teaching around critical literacy, where a change in consciousness comes about when the people feel themselves empowered to help bring about that change. In another example, The Pacific Northwest Earth Institute, a group representative of the environmental ethic of the Northwest, teaches deep ecology, practical eco-centered workplace practices, and, beyond that, provides an extended, ecology-based personal community for individuals and businesses. The goal of such a program is, again, to create a comprehensive cultural shift in consciousness.

The programs in the Northwest have historically included a blend of social, academic and organizational innovation combined with deeply held internal and spiritual beliefs. In the Northwest, a site of geographical exploration of land and water, students in cities now learn the value of natural resources as elemental to their own survival. In urban gardens, students are invited to engage in the development of their own urban ecological literacy; to simply sit and take note of the natural environment that supports life itself.

Urban Environmental Northwest

> The crisis…is not in land use. It's in the lives and minds of land users. That's why I don't believe it can be helped very much by any official policy. Good land use is going to come either by hard necessity or by some kind of teaching. (Berry cited in Musick 2009)

At the turn of the last century, the notion of a garden paradise based on a rural ideal formed the basis of many urban school gardens. Now, a less literal symbol of the natural world, popular in the Northwest, represents a garden as cultural tradition that is at once community-based, vernacular and sacred. A garden as sacred space endures and remains unthreatened because its ideals have permeated culture. A Northwest or "Cascadian" perspective on nature can thus be interpreted in practical terms, as a way "….to engage the natural world with wisdom and maturity, not merely romance. Can Cascadians learn to revere and restore their natural environment while using it as the vital resource base that it is? There are hopeful signs throughout the region." (Shibley 2008, 49) A vision of sustainability in the region includes practical application of values even as the application of values influences practice. In such a vision an urban school garden becomes one of those "hopeful signs." Another example would be a non-profit community organization like Sustainable Northwest, which emphasizes the need for practice in the context of higher ideals, in a society where people, communities and businesses work together for the good of the land, with recognition of and respect for the limits of earth's ability to provide. Here, a cultural concept of the environment is also influenced by a symbol of wilderness, what one Canadian pollster calls an "elusive utopia…of what is, and was and could be." (Greenville 2008, 53)

In the Northwest, a shift in values is thus not only an idealistic goal but is seen as part of practical and cultural change. The idea of nurturing values in practice (and practice in vision) is a theme that, here, helps nurture a new well-being in gardens, cities, and economies. Several different perspectives and organizations have contributed to this unique spirit of place. Two exemplary programs in the Seattle area have particularly helped situate the urban school garden in local grass roots practices and regional philosophical traditions: Seattle Tilth and P-Patch Community Gardens.

Seattle Tilth

Tilth of Seattle started in 1978 with a goal "to transform the Northwest region into the organic gardening capital of the world." (Musick 2009) The original Tilth garden began with a project at the Good Shepherd Center for the specific purpose of developing a unique urban agriculture centre. Tilth volunteers began by replacing the concrete sports yard and building a garden that would promote farming education as well as inspiration. Today, their work includes demonstrations, year-round vegetable gardening, fruit trees, permaculture, rain water harvesting, green roofing, composting and soil building programs. The Children's Garden provides a gardening education program and camp for approximately 5000 people each year.

From the start, the project grew out of difference, and evolved in a way that was typically counter to mainstream culture. While the garden began in 1978, the origin of Tilth goes back to 1974, when a loose regional network of organic farmers and gardeners in the Pacific Northwest attended a symposium in Spokane entitled "Agriculture for a Small Planet." Here, Kentucky farmer, poet and writer Wendell Berry was said to have spoken forcefully about the roots and culture of agriculture and the belief system that inspires organic farming and community. Based on Berry's challenge, the Northwest Conference on Alternative Agriculture took place later that same year as a combined gathering of farmers, educators and alternative thinkers; such a gathering had not been seen since the turn of the 20th century.

A new agrarian coalition, a Tilth Association, was begun. Through Berry's initial ideas and the continued work of Tilth, food and farming was viewed as integral to culture. As Berry stated: "A culture is not a collection of relics or ornaments, but a practical necessity, and its destruction invites calamity." (Berry cited in Music 2009) His words brought a new imperative to support the earth—as food, as farm and, as earth.

P-Patch Community Gardens

P-Patch is another uniquely urban community-garden based program with origins in the Northwest. The Seattle Department of Neighborhoods P-Patch Program now supports "P-Patch" community gardening, market gardening, youth gardening, and community food security in the City of Seattle. The program is for all of Seattle, but the unique emphasis of the program is on low-income and immigrant populations and youth; the program sends 7 to 10 tons of produce to Seattle food banks each year. The P-Patch currently supports 1900 plot holders and more than 3800 urban gardeners on 23 acres of land.

P-Patch also began as a community-based agrarian ideal, supportive of organic gardening, and volunteer community membership. (City of Seattle Website) The history of P-Patch is true to the region: an informal gesture of generosity and good faith filled a community need then blossomed into a thriving, sustainable environmental network. (Hucka 2009) In the early 1970s, Seattle farmer Raine Picardo (where P-Patch gets its name) lent his unused family farm for a community garden. Rainie's advice to all who took part was: "You get back what you put in." The origin of the program started in 1970, with the "Boeing Bust," when the company was laying off workers who suddenly needed jobs, money and food. At the same time, social activism was on the rise and the public was moving back to the land. A student at the University of Washington named Darlyn Rundberg Del Boca, was inspired to bring children in the city to grow food in a program called Neighbors in Need, a forerunner of the Food Bank program. The Picardo family farm gave permission for the group to use a portion of the land in Northeast Seattle, and worked with the city on a program where the land could be leased for the price of taxes. (Hucka 2009)

In 1973 the City of Seattle bought the Picardo Farm property and in 1974 authorized a community gardening program to promote recreation and open space. The program expanded into a number of community gardens built throughout the city, and was named the P-Patch Program. A major expansion in the 1990s saw a group called Cultivating Communities, a collaboration of the P-Patch, the Trust and the Seattle Housing authority, began to focus on gardens for low-income housing in Southeast and West Seattle. The evolution of

P-Patch speaks to both the city of Seattle and to the Northwest region; from the initial donation of the P-Patch, through to the grass roots support, donations and expansion, Seattle began to become known as a centre for thriving urban community gardens.

Maple Elementary School: Seattle, Washington

Places in the imagination that have a kernel of reality always possess an economy. (Wexler 2008, 215)

Maple Elementary School is in the Beacon Hill neighborhood of South East Seattle. Maple has more open land than most other schools in the area, and it sits on a former wetland. The garden at Maple is thus a rarity and an important part of the urban ecology of both the neighborhood and the region.

A painted mosaic, created by students, sits mounted on the brick wall outside the front door of the school. The mosaic is a blend of student-made artifacts that symbolize diversity in society and nature. The piece symbolizes how small-scale, cultural symbols take place both in their own right and as a larger vision of an urban public institution.

The school is located atop Beacon Hill, with views of the Olympic Mountains and Mount Ranier, covered in snow. The school has 60% of students on free- or reduced-lunch programs; the school has a high ESL population, a Behavior Intervention Program, and is also on par for standard test scores. The Maple Creed reads: " I know that our country was built by people of all races, and I know that people of all races keep our country great."

Near the front of the school a mural entitled "Volcanoes," painted by Mexican artist Victor Ayotl depicts two mountains, the Ixtaziuatl and the Popokatepetl, on either side of a river. The sun rises over rainbows, children, warriors and the village below. The image contains a blend of humanity, nature and civilization living together in co-operation and harmony. Small children climb the rainbow and swim in the river, below the two mountains.

On the other side of the school, the Maple Elementary garden sits beside a brick building, on a rise. The space itself is not large but certain design elements were used to enhance the garden and give a feeling of open space. A wooden archway leads into the garden, which is a re-creation of a small farm, with a greenhouse, a small barn, bird houses, a rain barrel, and hand-made fencing. On the other side of the garden, between the garden and the brick wall, a gateway leads to an empty grassy field that gives the garden a feeling of a wide

FIGURE 12. Students work in the Maple Elementary School Garden, Seattle.

open, rural field. In fact, Interstate 5, the railroad, and SeaTac airport are all located just below.

The garden teacher, Marcia Ingersolv, works with her students to grow a variety of plants every year, that include ginger, carrots, goji plants and a tall eucalyptus tree, which sways in the breeze above us. Marcia explains how this year she grew plants that represented settlers who arrived from different parts of the country, and the students researched and planted what the settlers planted as part of a curricular unit.

Marcia has involved the students in harvesting seeds from plants as part of a science unit. From there, students decided to start a seed company, where they designed the package, wrote the instructions and packaged seeds for sale. Heirloom tomatoes, basil and other herbs grow in a greenhouse that Marcia won for the students in 2005. Marcia also involved the students in making and selling earrings, which helped raise money for the garden. The class earned enough to build materials to build a barn. The long-term goal is to earn enough money to create an outdoor, science learning center to support the science units taught at the school throughout K-5. The longer-range plan is to include a bog garden that will hold the runoff and help re-establish native plants to the native wetland; there are also plans to develop a learning pavilion (with a living roof) that will promote and model ecological design methods of construction. The garden is designed around a theme of economic self-sufficiency, shared work, and community, while also leaving room for the many undiscovered, ethereal elements of nature. (The Maple garden also has a small "poet-tree" and an "elf-chair.")

The area was at one time all asphalt, before Marcia created flat garden beds with (found) wood dividers. As Marcia explains, the students used to play "wall ball" there. When the asphalt was removed, the soil underneath was "hard as rock." Marcia and her students began with a plastic greenhouse that they hammered into the rock. In 2006, a major windstorm unhinged the greenhouse, and it rolled down the street. She points to the sturdy glass replacement greenhouse they won in a contest. As part of the grant, the students wrote about their experience growing a beanstalk that somehow grew (and grew) in an unheated portable classroom. As Marcia explains:

> . . . the plant kept growing and growing. And it grew for almost the whole school year. We wrote stories about the magic bean, the plant that survived without sunlight. It was an indoor beanstalk! Every student had a different story.....winning the grant was a combination of creativity, and need.

The students also saw the destruction of the greenhouse as a way of impeding not only their learning in the garden but also their small economy that depends on growing/selling plants. The students understood implicitly the connection between the land, the tool for their work, the small-scale 'market' economy of the garden, and the survival of the site itself.

Community Connections

Having students write about the garden not only helped them fund their garden, but also was a way of bringing together the diverse students who arrive at the school. Marcia explains that most of the students at the school are originally from countries with strong agrarian societies. Since many families live in apartments, the garden thus provides an opportunity for parents and the larger community to take part. Thus, every student and family brings their own story to the garden, from their own, diverse perspective. One tree had a profound influence on one of the parents, as Marcia describes:

> This tree is a eucalyptus and I was out here with an African parent one day, from Somalia, and he looked at the tree and said: 'In our country, when we see one of these trees, we know there's water. We have a visceral reaction to that tree, a happy feeling of relaxation. When you see that tree you know—water is here.' He said 'when people in my country see that tree…they feel it.'

Marcia explains that in the context of the parent's story, the tree is now a symbol of "serendipity" that allows the community to bring diverse cultural meaning: "Someone happened to give it to us. This garden means so much to so many people."

The students have also used the garden in tandem with a novel about elves where, like the character in the book, they built "elf furniture." Marcia observes how an outdoor classroom brings a site for creative experimentation, where curriculum comes to life. In addition to building areas in the space that stimulate imagination and empathy, the garden also contains a solar-paneled birdhouse, built by one of the VIP students. In all cases the students make decisions around the garden, collectively decide how the money they earn will be spent, and whether it will be donated or used for the garden. Every year, Marcia's class is also involved in a service project connected to the garden. One year, from the garden earnings, the class was able to finance a water buffalo for a farmer in Vietnam. This came about from student research, where they discovered there are approximately 185,000 unexploded land mines in Vietnam. The students brought this information to Marcia and she responded by asking them:

> Now that we know this information, what do we do about it? And because we are farmers we connected to a farmer in Vietnam, he told us what he needed, and so we raised the money to send him a water buffalo.

Marcia adds that the value of the school garden for the students in this neighborhood is "undeniable" as it brings motivation to learn, and health:

> One year we grew spaghetti squash…so we roasted the squash, along with tomatoes and carrots, and ate it in class. The next day I had parents come in, all asking: what is spaghetti squash? Soon, they were all making spaghetti squash at home.

Marcia says that the garden "opens people up to new vegetables, new food" that invites people out of their usual patterns. She designed the garden with a multi-year curriculum in Earth Science in mind, with full support from the principal for an enhanced garden-based-curriculum. The school efficiently uses the garden; all classes use the space. The garden is directly outside Marcia's classroom and here she now explores Earth Science through plant growth and development for Grade 5. Each grade level has an Earth Science unit that brings students to the garden. The themes are: Kindergarten: animals; Grade 1: organisms; Grade 2: soil; Grade 3: plants. Grade 4: eco-systems; Grade 5: land and water. Marcia says, "We are creating a space where the students can experience the concepts embedded in the units first hand." The garden is used as a place where students go when their work is finished; the garden becomes their reward.

The garden is now in its third year and, over time, more and more students and teachers use the garden. The extension of the cultural components of the garden to the rest of the school community are important in helping the garden endure.

As part of Marcia's well-planned design, students enter the garden through a cedar trellis (also won through a funding competition):

> I decided the garden needed to have a secret entry way....a special way to enter the garden. The students call this the "The Secret Garden."

On whether or not teaching in the garden is organized or spontaneous, Marcia states that in fact the key to being able to be spontaneous in the garden arrives when teachers have access to a wide range of curriculum materials. This way, teachers feel free to spend time exploring the garden itself as each lesson need not be developed from scratch. For Marcia at Maple, the curriculum grounds the lesson, allowing spontaneous learning to take place. In encouraging other teachers to use the garden, Marcia takes a relaxed approach that especially involves respect for time constraints:

> It has to be a bit *catch as catch can*. If you tell people it has to be a certain way, it will just become another job. Spontaneity is all important; the sidetracks are important. For example, when we go out here we have teams: *weeders*, *waterers*, and *worm wranglers*. We have stations out here and students get really invested. The jobs rotate. And they have their role. So we try to make that part as easy as possible for the other teachers to facilitate.

In connecting teaching in the garden to curriculum standards, Marcia sees the value in simply giving teachers a starting point. On a tour of the garden, Marcia points out the Three Sisters garden she uses with a Grade Three class; when this part of the garden gets taller, it will also help shade the greenhouse.

FIGURE 13. Students working in the Maple Elementary School Garden, Seattle.

From design to curriculum to practical farm knowledge to creative learning, the entire garden is valued, planned and organically connected as a functioning system. Marcia explains how the cycles of the garden inspire learning as process and community: "When the grade 4 group comes out they relate to the Three Sisters they worked on the year before. It's another way of bridging grade levels."

The school's VIP Program (or Behavioral Intervention Program) is also connected to the garden. As Marcia explains, students in the program have become "champions of the garden. The kids are so focused here, they work tirelessly. You can have kids with behavioral problems and they come out here…the transformation is incredible."

The "poet-tree" planted by students is in fact a small madrona tree (a stage) from a cutting from Marcia's backyard. The kids sit on the wall, and on the steps, and recite poetry here. The poetry program is successful and inspires "magical" learning and inspiration, especially for those who need it most. Marcia describes the response to the garden from the Grade 1 teacher: "…she had kids who were normally uncommunicative in the classroom…who could suddenly speak out here. It's exciting to see. Students might not participate in the classroom, but when they come out to the poet tree….."

Marcia explains how the garden uniquely helps students in the urban neighborhood:

> We work with the students right where they're coming from. This is South Seattle. Many of the parents who come here already come from agrarian societies. They see the value in the garden right away. We take a 'scientific viewpoint'; even more parents in this area, with their backgrounds in agriculture, are going to be excited about it.

She adds that science projects in the garden usually arise from within nature's process:

> What they [students and parents] see here…they've already seen. This kale for instance. It was pollinated then we let it go to seed. Now we're watching the process. And, there's a control group. We didn't pollinate the collard greens over there, we let the bees do their thing. The students then note the difference. So these are budding scientists.

In the garden, students learn the basis of human survival through connection to land, food, and through the practical know-how of a small-scale sustainable market. From observing nature's processes as Science, the students then branch out to poetry readings, to global awareness, and to the creation of a seed company, where they learn marketing, accounting and business skills. She showed me rows and rows of heirloom tomato plants. The seeds carry over year to year, providing the garden with plants and seeds for the seed company. The seed in the garden is thus both a practical means of developing local sustainability and a low-cost way of expanding democratic participation.

The students at Maple are very involved in decision making (i.e., choosing what to sell and planning the garden year) and Marcia explains this is one way the garden directly invites engagement with learning. For example, one student who had been previously disengaged in classroom learning pointed out that they needed to label the plants or else "that chive or spinach and those things that we know need to be harvested will go to seed." In another example, students collaborated around lessons on the Oregon Trail unit, where students were grouped into families who settled on the Oregon Trail. In the garden, they now call themselves by their Oregon Trail names. Marcia pointed out piles of farm clothes that the students wear in the garden; these include denim overalls, boots and straw hats. Marcia explains that for students who own one pair of pants, having extra clothes and shelter from the sun gives students a further sense of security.

When the students researched and planted the plants used by pioneers, Marcia describes how students began to feel connected to both the land and to previous Northwest immigrant and settlers who struggled to survive. Marcia sees the garden as an important way for students to connect to the planet in an experiential understanding of history and time. Here, students consider how their work connects to a much longer timeline of shared history.

As the discussion ends, a Boeing 747 appeared in what seems like a very short distance above us. Marcia says that indeed the planes are so close "you can see the pilot's face!" Transported to the small urban farm at Maple Elementary, it was easy to forget that the airport, the interstate and the railway tracks were just the other side of the hill.

An Intentional Community

The teaching assistant at Maple Elementary, Ydalia Yado, calls the garden, "a beautiful place, an intentional community." She described the importance of students being able to bring home food grown in the garden and believes this has an enormous impact on the students because "if they see a plant in this small space they know they can grow it at home." Ydalia suggests that students in the garden create a sense of unity for themselves that makes their own community more visible:

> Now, in their own neighborhood, they are also seeing themselves; they work together intentionally to sustain themselves because they know after these plants grow they can eat them.

Ydalia adds that the value of the urban school garden is "its own living proof." The work in the urban garden is connected to people in the struggle to survive, worldwide, not always working in ideal conditions.

In the end, students and educators at Maple understand that a garden is work. And, the experience of growing food is directly related to survival and making a living. As Ydalia states, such understanding helps students contextualize their work in a global sense:

> In the garden they see how millions of people around the world make a living. Whether it's migrant workers, or farm workers....they know the actual work involved. When they see the food at the grocery store, they'll know it takes a lot of work to plant and grow these fruits and these vegetables. They'll know what it takes.

Students at Maple Elementary

At Maple Elementary, students shared photos of the garden and took pride in marking their progress.

Students grow, harvest and cook spaghetti squash, peas, pumpkins, onions tomatoes, potatoes, and carrots. Nutrition lessons focus on the sun's role in the lifecycles of plants.

Exploring student attitudes toward the garden, the teacher, Marcia, and the entire ethic of care and involvement at the school have helped make health and nutrition personally valuable to them. The Grade Four students sat in the garden, responded to questions, while jumbo jets overhead roared onto Boeing Field.

What do you like best about the garden?

Here, students described the various ways they are directly involved with the garden. Their hard work in the garden, as described by the garden educators, has brought a direct emotional connection. Their responses to the little patch of land outside their classroom door emerged lyrically, as poetry.

> *Best is learning the plants, and saying:* Hey, how do you do?

> *You harvest seeds. You know what plants are plants, what are weeds. And much much much much more. (Oh, so much fun)*

> *The garden gives you a great variety of ways to like to work in the garden and to not be afraid to get dirty.*

Several students responded to the ways the garden helps them address their own personal fears and brings them courage. They also uniquely recognized how the garden helped their own ideas come to life:

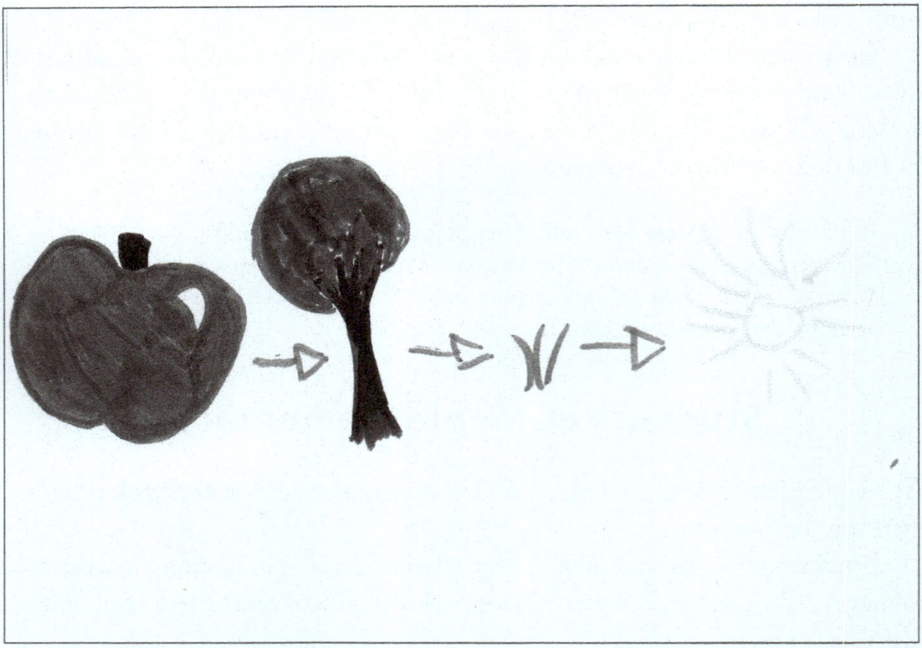

FIGURE 14. Students depict the source of their food connected to the earth and sun.

Out there you can face your fears. (If you're scared of spiders or worms or anything)

I say take a chance, try everything you believe in. (You can be un-scared of anything)

Right now we're just speaking of stuff. (But it is still true)

The students revealed how the garden changes their sense of space, through observation of the plant cycles:

The special and cool thing I learned in this garden is how the plants grow

You see the environment grows into what is around it and what places the plants need to grow into.

While other students responded:

The life of the plant? It's all full of life!

It's the best school with the best teachers and we want to keep it nice and healthy. And cool.

Here, the garden is viewed as a small, shared economy:

> *Look—even if you have a store with no corn you can always ask your friend: "some corn seed please."*
>
> *Sharing is the key.*

Students described how the most important aspect of the garden was the diversity of life found there:

> *You know how you think spiders and worms do not have a life or a heart? Actually they have a heart and they have a life and they might be smaller or different from us but they still have dignity and life. (So never kill any life)*
>
> *Everything is important in the garden.*
>
> *Yes, pretty. (But weeds in between)*
>
> *Well water spills everywhere.*

Describe the difference between learning inside and outside:

Here as elsewhere students related the garden to direct, sensory experience and to feeling more connected to their local surroundings:

> *Outside in the garden you can see what is going on and all of the things that are happening around you.*
>
> *Outside. You have more experience outside.*
>
> *You can see out in the garden.*
>
> *You see how it kind of shines if it's watered and if it's wet you see how the leaves shine.*

The students felt that in the garden they could reveal their differences and their inner life:

> *Out here I could feel like super heroes.*
>
> *Out here....I got the uniqueness.*

What would you like to tell the world about your garden?

Here, students imaginatively described their sense of the space. The garden represented the larger world and thus, learning in the garden transported them to other places:

It's like seeing the whole world seeing the insects. If you see everything here you see everything everywhere.

If you see insects here you feel like you're in a safari zone. You just feel happy. You don't feel scared. You just feel calm.

If you wanted to see some plants, if you were anywhere in the whole world, you'd probably want to come here.

A sense of connection was inspired by the place and extended to connections between the students themselves:

We even do our own science out here. You're with your friends. People you know. People you know about.

The students gained a sense of empowerment and joy from being in the space itself:

If you're there you just feel happier. Cuz if you're inside it's all bricks and math.

Here, you walk around and see different stuff in different spots. (There's a spider! There's the food!)

The is the best spot, just look at the plants.

Conclusion

here is the root of the root and the bud of the bud
and the sky of the sky of a tree called life; which grows . . .
(e.e. cummings)

In the context of the Northwest the garden at Maple Elementary reflects a theme of connections: between people, land, learning, time and process. In observing the work at the garden, there was a sense that even the most practical actions were guided by deeper beliefs and that somehow these beliefs had been made real through the site itself.

The garden at Maple begins with the mosaic at the entrance to the school and in the artist's mural inside on the plain brick wall, depicting the Ixtaziuatl and the Popokatepetl mountains. The theme here is dedication, loyalty, and practicality, blended closely with humanity, ecology, and story. The idea of humanity in harmony, connected to a much larger narrative, is a theme of the school in both the context of the Northwest and in the teacher/student stories. Their words, actions and beliefs are integrated into the school and garden in a way that humanizes the learning process. It seemed effortless.

The deeper themes of the garden are inspired by the work of leaders (teachers/administration, especially teacher Marcia Ingersolv and school principal Patricia Hunter) who encourage a wide-lens curriculum, experiential learning and student involvement, on a daily basis. At the urban garden at Maple students are true collaborators with their leaders and with land.

A cultural awareness of land as involvement, practice and progress emerges from long-held beliefs in the region. A recognition of values underpins the practical work while seemingly smaller tasks take place in a process of a larger perspective. Here, people seem to be aware of the scale of the work. The student work is imaginative (i.e., poet-tree, elf garden, solar paneled birdhouse); experiential (integrated garden-based curriculum at every grade level; daily use); and practical (the seed company, the greenhouse, the water barrels)

The barn, the native plants, the mural, the mosaic, the emphasis on diversity, the trees of cultural significance, the development of grass-roots local market "green economy" are also vivid examples of the emergence of a larger, self-sufficient community identity. The experience of new immigrants to the area is blended with a deep and authentic commitment to diversity, each other and to the planet. The garden gives the students a shared sense of identity and commitment that allows them to help others in a global sense; as their teacher explained, when students donated the water buffalo to the farmer in Vietnam, they were responding to a crisis—as farmers.

The responses also show how students in an urban setting value the ability to situate and transfer their knowledge; they have grown to love the plants, the garden itself, and the ability to have valuable encounters with peers. The vegetables grown from the garden and cooked in the classroom have become part of the students home gardens and home cooking. They show a unique awareness that this site, with its insects and other flora and fauna is connected to the world, and also represents the faraway "safari place." Practical knowledge of the garden gives students solid understanding of their role as growers and, with such knowledge, they develop courage in being able to face their fears, and become "unscared." They described their new and vital awareness of gardens as a life cycle that always brings a timeline for growth, change and a sense of hope. The garden gives them a renewed sense of place, as they saw plants in the world "in a space" but also a space "that is moved into." They demonstrated their knowledge of the garden as a place in the school, in the neighborhood, in the city and beyond. The learning in their "Secret Garden" is a portal to the world and to their imaginative re-creation of it.

Knowledge, combined with practice, brings empowerment. A common value observed in all of the students at the school was the importance of "staying real." Being able to record their ideas and talk about the garden as "real" also made their ideas "true." The students were very aware that their work and their ideas were "made real" in the garden. They saw themselves as transformers who interpreted their work in the land into idea and story. The urban school garden visited in the Northwest was a living place that seemed to demonstrate students' understanding of both the practical and profound elements in nature.

In the larger story of the Northwest and its history of progress and vision, such initiatives seem to arrive spontaneously. But the actual practice of the work is firmly rooted in traditional community values of cooperation, learning and spirit. For students in an urban garden learning brings survival. Here, they learn "what it takes."

· 6 ·

WEST COAST, CANADA

Garden as Community

Introduction

> The water, it's a gift of life. It bothers me because our water is...disappearing because it's not being respected.
>
> (MARY LOUIE, OKANAGAN ELDER, CITED IN BLACKSTOCK 2001, 4)

In a traditional indigenous worldview, support for the environment is rooted in respect where "water can be given and it can be withdrawn by the 'very something' that creates and sustains all life." (cited in Blackstock 2001, 4) In cities, the notion of sustainability is always open to cultural interpretation, to a widely diverse population's worldviews and ideas. If respecting and valuing nature is central to sustainable culture as a guiding principle, that value is even more necessary to the day-to-day life in dense urban and suburban areas.

An urban school garden in the setting of Vancouver, British Columbia, Canada, promotes both the idea and the experience of a diverse culture of sustainability. In the same way that natural resources that are commonly respected are also sustained, urban ecology depends on diverse perspectives that both embody, and nurture, the process of environmental learning. Simply put, discussions around sustainability need also be sustainable, and earth centred.

The garden described in this chapter provides an overview of how students and teachers in urban school settings meet the natural environment in small, important ways. The garden here is a symbol, and a microcosm, that reflects the proximity of learners and urban nature. A school garden surrounded by urban development becomes a symbol for sustainable teaching and learning, rooted in the land. In such a setting, a school garden becomes a blend of multiple perspectives involving teachers, students, worldview, location, learning, curriculum and the context of the city that surrounds it.

The story in Vancouver begins quite differently from the other gardens described in this book. Instead of starting with the garden then moving toward cultural context and symbol, research around the urban school garden in this area began with the students themselves, then moved to the garden and local landscape, where their firsthand accounts became creative expressions of experience. Instead of the garden leading to student expression, in this case, student response to their local surroundings, lead to the garden.

Canadian Context

In ways that are similar to the United States, urban development in Canada incorporated school gardens early on, with an emphasis on practical training. Agriculture was taught in schools in 1872 and by 1899 programmes thrived due to the incorporation of agriculture courses in teacher training. By 1918, every province in Canada promoted school gardens; after WWII, Victory Gardens continued to thrive. (Davis 1903)

Canada's urban school gardens were also promoted by the Commissioner of Agriculture and Dairying and the superintendents of Education, in several Eastern/Maritime provinces. The Macdonald fund, or Macdonald schools, thrived in Ontario and in most cases, agriculture was taught to urban youth as a trade skill. The focus turned to the large swathes of remote, empty rural farmland and urban schools became training grounds for future rural farmers. Throughout most of the 20th century, the focus in Canada was aimed squarely at economic, agricultural and training goals.

Environmental Education in Canada has thus evolved from a strictly utilitarian perspective in the late 19th century to an embrace of the deeper goals and visions that might guide learning for sustainability. Today, the ministries of Education in Ontario and British Columbia have adopted environmental education-based curriculum guides that move well beyond training goals. The guides promote broadly defined notions of an ecology-based pedagogy that includes student connection with nature. For example, The Toronto District School Board uses three broad concepts in integrating environment to established curriculum:

1. Sense of Place: to explore and see our immediate surroundings, both natural and built, and to look beyond the immediate to the larger landscape.
2. Ecosystems Thinking: to help us examine how nature works and to understand how natural and human systems are interconnected and interdependent.
3. Human Impact: to weigh the consequences, both helpful and harmful, of human interventions into natural systems. (cited in Houghton, 20–21)

The B.C. Ministry of Education curriculum supplement: *Environmental Learning and Experience (ELE): An Interdisciplinary Guide for Teachers* (2007) designs curriculum around the following principles:

- Encourage the integration of subjects/multidisciplinary approaches
- Encourage critical reflection on a range of perspectives
- Examine issues for their currency and authenticity
- Acknowledge aboriginal perspectives
- Acknowledge other perspectives
- Consider the place of action
- Consider issues from both local and global perspectives
- Occur within a context of hope
- Encourage humility

(B.C. Ministry of Education 2007)

Unique to both of the guides is the departure from a how-to model of outdoor learning. The new curriculum documents encourage learning in and about the land in a way that promotes personal transformation. The use of words like "hope," "humility," "interconnected and interdependent" are strongly influenced by global environmental research from the UN that emphasizes how the effective integration of ecology in schooling must begin with high ideals, broad visions, and hopeful perspectives as integral to a sustainability society. Such a vision uniquely relates to the complexities and multiple approaches to sustainability required in cities where, as elsewhere, challenges run deep.

In urban Canada, school and community gardening has followed a rise in popularity similar to the United States with roots in the late 19th century, followed by strong institutional support in the early 20th century, and a steep rise in interest in the mid-1990s. The Toronto-based non-profit group Evergreen, a supporter of school and community gardens, studied and recorded the experience of six schools involved in a school ground naturalization project. The study examined the outcomes of school garden transformation, where asphalt and grass has been replaced by native plant restoration. They noted a summary of outcomes, namely, that knowledge related to ecological literacy improved, and that students gained a stronger sense of school community. In Vancouver, in particular, the interest in school gardens has dramatically risen in recent years, and the provincially funded school district is beyond capacity to provide the schools with personnel to support the strong interest in environmental programmes.

With the Vancouver and Toronto Board of Education's leading curricular goals of developing a sense of place, provincial curriculum standards now support more open-ended values in learning in nature. In Canada, the development of urban school garden culture has evolved to a balance point where philosophy and ideals guide practice, and practice lies at the centre of pedagogy.

In Practice

Projects in the region describe a culture that includes a history of strong grass roots practice combined with deeper ecological ideals. Three main Canadian organizations (City Farmer, Evergreen and *Green Teacher*) have worked consistently to promote practice and research in school gardens across the country, either through publications, research funding, or workshops. City Farmer is a comprehensive resource for urban community and school gardening. The group provides lessons in organic gardening and worm composting through a demonstration garden, while the website Urban Agricultural Notes provides an important record of first hand accounts of practice for the region.

In Vancouver, the urban school garden movement largely emerged spontaneously through word of mouth and community-based movements, often ignited by the passions and imaginations of individuals, volunteers and professionals. In Vancouver, along with civic and community networks, university researchers were involved in the documentation and conceptualization of an urban school garden culture from the start. *Stewards and Storytellers: The Greening of British Columbia School Grounds* (Evergreen 2000) describes practices of creating organic links between learning, land and narrative. For example, Lord Strathcona Elementary School is an inner-city school that developed a naturalization project in the midst of downtown apartment blocks, surrounded by high crime statistics. The school coordinator described how the garden evolved from student interest in urban nature: "For years before we put the garden in, the kids would dig their own streams and things...they dug their hands in, taking greenery and sticks and making their own little world."(29) The Vancouver School District was involved in providing maintenance and grounds crews.

As in any community, a challenge for the urban school garden involved logistics, safety, maintenance and the need to convince a community that a garden held more practical benefits than a flat playing field. The neighborhood presented its own challenges: "Anytime kids will stick their hands in dirt, there's a concern that they'll stick their hands on needles...." And yet, a commitment to local environmental design, responsive to local landscape and respectful of waterways prevailed.

One of the first school gardens in Vancouver (established in the mid-1980s), the garden at Lord Roberts Elementary in the Downtown West End has grown food in the city and taught students about nature, from within urban nature. The project was created during a time of increased conservatism and demand for strict guidelines around provincial standardization and testing. However, the garden at Lord Roberts was well known for teaching students about the land and about values related to empowerment, community and hope. The project now documents the origins of an elusive urban school garden *culture* in the making.

The project involved 250 students who learned about connections to the environment through planting and harvesting vegetables. One goal was to involve students in beautifying their concrete neighborhood while gaining a new understanding and respect for the miracles of nature. The garden pedagogy newly focused on student connection to urban neighborhood and to their understanding of the connections between learner, nature, and city.

The project was lead by Dr. Gary Pennington, a UBC professor of Education, who co-ordinated the project and found approval from the Vancouver School Board. Pennington noted how "The food garden increases their sense of wonder. It's pretty magical, but at the same time, it demystifies the concept of food production." The project thus found leadership in academia, combined with support from City Farmer's expertise in community organizing. The result is an urban school garden that exists to this day in the centre of downtown Vancouver, on the ocean's shore, amidst high-rises, hospitals, and commercial zoning.

Interestingly, and uniquely, the naturalization project at Lord Roberts was built, from the start, with notions of transformation of urban learning. As the garden's founder (Pennington 1988) states: "What we see and understand about the playground is governed by the way we look at things." Pennington describes how the space teaches a new kind of awareness, evidenced by: "Children disproving adult cynicism by nurturing rather than trampling small flowers and shrubs which they have planted with their own hands…." and "Kids learning about the good earth, that fruits and vegetables don't necessarily come in cans, and that we can grow our own food and flowers." Pennington stood firm in his advocacy for promoting "the human side of learning."

> It is obvious to those of us who are intimately involved in the project that the human side of our enterprise is paramount. An increased environmental awareness on the part of all is emerging….put very simply every time a young child puts his or her hands to work with the garden trowel, shovel or paint brush, there is a major chance for learning and a much reduced chance that alienation, vandalism and violence will occur. (Pennington)

Urban school gardens in the Vancouver area involve practice, philosophy and a sense of transformation that arrives through deceptively simple goals. The educational philosophy base describes "simple yet profound principles that converge when we discuss children and gardening." (Ibid.) Based on his experience at Lord Roberts, Pennington describes how transformation is a curricular goal and a potential outcome, rooted in the core values of respect, community and diversity:

> Gardening can be a transforming activity for young and old. It can move us from ignorance to understanding and appreciation, from passivity to action, from consumption to production, from silence to dialogue, and perhaps most importantly, from a state of dependence to one of independence with nature and others in our community.

Grandview/Uuqinak'uuh Garden: Design and Practice

> Freedom has many exotic colours
> that light up the darkness
> (April Hale)

The garden at Grandview/Uuqinak'uuh Elementary School is located in East Vancouver, near the Downtown Eastside, the lowest income neighbourhood in the country. The school has a high aboriginal student population, as well as high populations of ESL students recently immigrated to Canada. In many ways the Grandview/Uuqinak'uuh garden, in its diversity, design and collaborative plan-

FIGURE 15. Cob toolshed. Grandview/Uuqinak'uuh Elementary School, Vancouver, Canada.

ning process exemplifies a uniquely Vancouver model of an urban school garden.

The garden began in fall 1998 through a meeting of the minds between the school and a university researcher. While school administrators were eager to improve the learning potential of the students, University of British Columbia Masters student Ilene Pevec was interested in conducting action research in outdoor learning environments at Grandview. Pevec's project incorporated nature, community and culture into a teaching curriculum, and she collaborated with landscape architecture student Tracy Penner, who would bring new ideas in sustainable design.

The original vision for the garden involved both university scholars and school administrators/teachers who held strong independent visions that coalesced into a community-based, collaborative action plan in an urban school garden. Their mission statement included ideals of community and sharing, in which a school located in a lower-income part of the city would provide a gathering place for learning, farming, food production and indigenous education. The garden would provide food resources for the community and "an enriched real-life curriculum" where "cultural references reflect the neighbourhood population." (Penner, 2000) The mission statement describes how The Grandview Community Schoolyard project would improve the quality of life for those living in the inner-city neighbourhood. Grandview would "also function as a living laboratory and a model of an urban ecological school yard." (Ibid.)

The space aimed to blend learning practice with the goals of the local community; the garden's ideals, in fact, grew out of both learning and life needs amongst the neighbourhood population. Early goals for the garden arrived through community consultation that involved students, teachers and local community members. Penner (Ibid.) describes how:

> Every child and staff member, and interested community members at Grandview have participated in envisioning and designing the school grounds to be a place of diverse natural life and beauty. Every child has grown plants from seed and planted them in the garden boxes built by the grade 7 students. Every child has eaten the tomatoes and picked the flowers they grew. The staff has formed a school garden committee; the community has formed a community garden committee.

Student surveys showed the vision of the students who wanted water somewhere in the garden, plus a hill to climb and flowers and berries. Teachers wanted a covered outdoor classroom for teaching students during the long West Coast rains. Meanwhile, neighbors desired a community gardening area for vegetables and flowers. Students were especially eager to involve themselves in the design process, and they always made special mention of water and sacred elements related to water:

> Almost every child in the school wanted running water and bridges. One little boy had me meet him on the playground so that he could show me where he wanted the waterfall to go. The children wanted ponds, streams and forts, tree houses and places to sit privately, places to hide..... one grade one student astonished me by creating a temple of shells on the top of a hill, a sacred space. (Pevec 2000)

They excavated a dissipation pond (one meter deep by 15 meters long by 10 meters wide), lined with large rocks on the bottom, smaller rocks and finally sand, to create a coastal beach front. One of the original plans included the idea of using crushed seashells around the outer edge. This plan was in honour of the Coast Salish, who have traditionally used midden shells to create and decorate pathways, many of which are still in existence hundreds of years after being created.

The pond met the challenges of the safety standards because only a few inches of standing water would collect after rainfall. The pond also absorbed excess water to keep it from city storm drains. At the suggestion of Grandview's teachers the garden was based on a First Nations design model and built around a traditional Coast Salish style Long House with Douglas-fir cross beams and red-cedar planks. The Long House, the selection of plants, in fact, the entire plan of the garden was all intended to honour the First Nations. The orientation of the Long House reflected the cardinal directions with the front door facing north and the rear door facing south; the building was situated next to the pond so that the entire space would reflect a West Coast fishing village. Students were involved in the symbolic decoration of the structure and assisted in painting the totem poles at the entrance. One class visited the B.C. Museum of Anthropology where students researched, recorded textile patterns and spoke with traditional weavers. The patterns were transferred onto graph paper and then formed the basis of the design for the patio bricks. (Pevec 2000)

Principles and strategies common to the development of an urban school garden included an emphasis on: a diverse cultural student population; a close neighbourhood/community connection; maximized use of space and room for uniqueness and spontaneity in urban learning. The public planning workshops noted the following points and recommendations that emerged from the planning process:

- First Nations culture is an important part of this neighbourhood.
- Lack of park space in this neighbourhood. (Parks board verifies this.)
- Need for more programs for parents and children of this neighbourhood, especially literacy and life skills.
- Desire for more community garden plots in this neighbourhood.
- Site currently has some problems with personal safety on weekends and evenings—vagrants and IV drug use and prostitution due to the lack of use and surveillance.
- Dog defecation on open field impedes use by children.
- Field and playgrounds suffer from poor drainage conditions. (Penner 2000)

Environmental design features in the garden were well planned in advance and divided into thematic sections, by use:

- Longhouse Outdoor Classroom: a community gathering place for outdoor learning and celebration. Design inspired by the Musqueam traditional longhouse;
- The School Garden Boxes: life sciences, applied mathematics, language arts, fine arts, multiculturalism, good health;
- The Natural Habitat for birds, butterflies, hummingbirds and other insects;
- Environmental Restoration, Natural Drainage: shaping one's environment to invite

wildlife back into the city, understanding natural functions of the environment;

- The Community Garden: families work together for the health and self-satisfaction of growing organic foods;
- Skills learned: gardening, nutrition, cooperation, cooking. Site safety improved by adult presence during hours when school not in session;
- The Ethnobotanical Garden: cultural and environmental restoration. The First Nations children and families will learn the native names and uses of local indigenous plants;
- Salish Patio and Drinking Fountain: a public art project created by youths and managed by local artists; water as the source of life in a place of beauty. (Penner 2000)

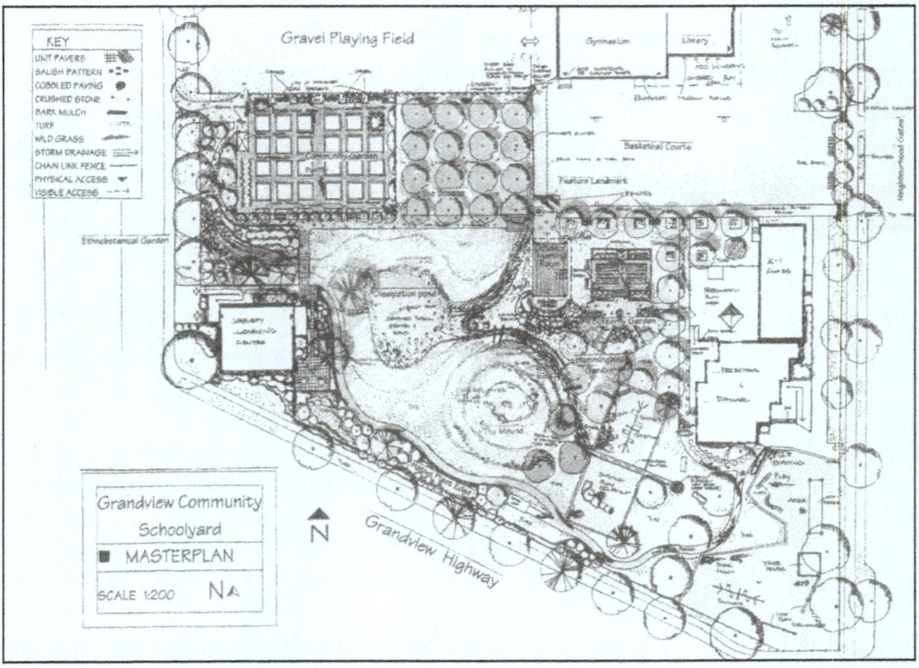

FIGURE 16. Grandview/ Uuqinak'uuh School Garden Design Plan. Tracy L. Penner, Landscape Architect.

A comprehensive curriculum was also developed for teachers in a publication entitled *Ethnobotany: Patterns in Relationships*, a curriculum for grades 4–7 distributed by Evergreen. (Pevec 2002) The curriculum in the guide describes ethnobotanic plants used traditionally by First Nations people. The main unifying theme for the Grandview/Uuqinak'uuh garden emerged from a goal of "reconnecting people to nature and nature to cities." (Penner, Ibid.) The design principle involved gathering resources from First Nations planning and construction techniques that also involved social design principles and ecological restoration. The design elements at the Grandview/Uuqinak'uuh garden also represent the theme, in its combination of hard landscape (deck, patio, cobbled walk, cedar plankway) and soft landscape (climbing plants, fruit trees, living fence, vine-covered walk, maple leaf bosque).

The long-term survival of the garden was also an important part of the original plan. The Environmental Youth Alliance (EYA) volunteered to work with school students to keep the native bird habitat maintained. An offer was made by the Vancouver School Board to prune the trees and shrubs on site with the high school students in its horticulture training program. (An innovative and badly needed form of support that, unfortunately, did not take place) Also in the plan was a proposal for a new part-time school district position of Garden Coordinator at Grandview, to be by the local public health agency. The position (still in place today) would ensure a legacy of encouragement and ongoing involvement of the school's classes, teachers, community gardeners and neighbors. A key feature of the original plan was the involvement of adults in the community garden, who also provide a daily example of ongoing stewardship for the school's students and teachers.

The project was well planned and in many ways the plan itself is still a model design for a multi-purpose, ethnobotanic, inner-city garden. The design itself is described as a dynamic convergence of like-minded people who seemed to just "show up" at the right time. One of the designers/educators of the garden immediately noticed the need for both planned and unplanned teaching in the garden:

> If we had several activities planned, we could go with whatever worked best for the children. We took notes of what the children said of the little gardens they constructed and I photographed every garden plan made. (Pevec 2000, 31)

Pevec also noted that students interpreted culture using natural materials and interaction while designers and educators are normally trained to incorporate culture from the outside in:

> In retrospect I see that the teachers wanted manifestations of culture and the children wanted nature and to be free to experience nature as they chose and with risk like water and high places to climb. (Ibid.)

The combination of planning between professionals, educators, community, and the learners themselves all provided important balance.

A hummingbird/butterfly garden was put in place, a wild bird habitat, and a native species garden. Students worked with the Environmental Youth Alliance and the Master Gardeners to eventually plant the ethnobotanical garden (the first of its kind in Canada), which included plants traditionally used in food, medicine and materials of First Nations. As Pevec describes:

> They were excited and keen to have an ethnobotanical garden growing at school where they could harvest plants and use them instead of just looking at them and talking about how they used to be used.... By seeing this urban native plant garden we all began to see how we could use such a garden for the education of the whole community at Grandview/Uuqinak'uuh. (Pevec 2000, 32)

The Vision

The origin of the physical space of the garden is only part of the story. Pevec (2000) describes how the garden grew from twin commitments: "a profound love of nature and a profound sense of social justice." (4) Like many founders of school gardens, Pevec's work was inspired by experiences in her own life. A visit to Brazil brought her in contact with a former student of John Dewey:

> A professor of education from the Federal University in Rio arranged for me to visit the Escola Parque of the Centro Carneiro Ribeiro, a model school in São Salvador, Bahia, created by the minister of education, Anísio Teixeira. The school is located in one of the poorest areas of the city, Liberdade, where many homes are simply salvaged tin and cardboard. The Escola Parque embodies Teixeira's philosophy of education. (5)

In this context, Texeira's vision for the school was: "The student will practice in the school community all that in the adult community of tomorrow he will have to be: student, worker, artist, athlete, community member, which is the same as to say useful, intelligent, responsible and happy." (Texeira cited in Pevec 6)

Pevec's research for her project at Grandview/Uuqinak'uuh similarly began with need in the neighborhood. She observed that in 1927, when the school was first built, that there was indeed an unobstructed "grand" view of the North Shore mountains from the ground level. In later decades, in addition to extremely busy traffic corridors, apartment buildings, light industry, warehouses and subsidized housing grew up around the school.

When Pevec first visited Grandview/Uuqinak'uuh, she heard the inner-city project teacher discuss the social and economic situation of the students in many East Side schools. He described poverty as "the common denominator," along with the usual inhibiting factors: "lack of access to the broader community, low self esteem for many children and their parents, a sense of powerlessness to change the situation." A main focus of the school day thus rested on student health and providing basic needs. As Pevec describes, the stance of a gardener-as-researcher was always central to the space; any innovations to the garden would have to include everyone in the community, and emerge organically. The garden thus began with a desire to create a project with inner-city school children and their community that would also function as a site for multi-disciplinary, creative exploration and learning. From a seed of an idea the garden grew both as a garden and as a community:

> What has developed has involved every child, teacher and administrator in a school and a day-care, parents, many members of the surrounding community and funders from far and wide. This project has taken on an organic life of its own in accordance with its nature as a garden, and the research has grown from my role as a gardener sowing seeds of community. (16)

The development and implementation of a truly community-based design led to further progressive practices in the garden that in many ways still leads the way in terms of environmental projects in the region and beyond. The project is an example of collaboration in action on a living site that adapts and changes, where the project itself leaves a legacy of collaboration and change: "We engaged in critical collaborative action research together, questioning assumptions about physical and social structure and curriculum together and acting to change them." (20) A project that began with social justice and values of equality, and a shared vision, resulted in a space that now represents authentic mutual respect that includes the work, the community, and the land, itself. In many ways the original design and design process at the Grandview/Uuqinak'uuh garden mirrored an ecological pattern, with a "web of people....just like the web of life." (41)

Current Grandview/Uuqinak'uuh school garden coordinator Brent Mansfield continues to carry out the vision of previous garden educators, planners and designers who originally developed the site. He describes the benefits of having a cultivated space replace an empty field in a dense urban area:

"The garden provides green space. That was part of the original plan. If you look at a map of the neighbourhood, there is almost no green space." Brent believes that the dual school/community garden at the site is unique and provides an excellent way to get the whole community involved in sustainable food practices:

> The unique thing about this site, is it provides both a community garden and a school garden....it helps promote the idea that school is not just a place that kids "go to but don't live in." In reality, they live in the community and the community is part of the school....So, let's use this beautiful space, intergenerationally. Let's bring elders, family, kids.

Brent's special interest in school gardens is the promotion of food and food-based education, through the hands-on learning practices of school gardens. He suggests that school gardens have far-reaching educational, health and community benefits that begin with food and lead to access and justice. As Brent suggests, "the garden is experiential but, it goes much further":

> There's no better way than gardens to help us teach what we need to teach, (and aren't) School boards promote healthy living—a garden helps us teach food. It's obvious! There is no better way to help students develop affinity with nature than through hands-on learning....If we can transform the way we eat, we can transform our health. Societies organize themselves in terms of relationship to the earth, but, also in terms of access and justice.

Similar to Marcia Ingersolv's comments in Seattle, some of the lessons that emerge from Grandview/Uuqinak'uuh garden relate to encouraging widespread teacher involvement, and to simplifying garden access for teachers. Brent supports the idea that teachers in fact require a great deal of freedom in how they approach the site. While Brent hopes that garden practices are never mandatory, or forced on teachers, he still hopes the garden-based movement continues to grow in Vancouver, and beyond:

> I would love to see a garden in every school. In every single way, curriculum could be worked into the garden. Science is an obvious one. But what about Geography? Poetry? Oral language? Gardens make learning meaningful, so you're going to have meaningful writing out there.

The development and evolution of the Grandview/Uuqinak'uuh garden originated as a process that involved both environmental, cultural and pedagogical revitalization. Like other urban school gardens, the site provides a base for gardening, for food cultivation, for cultural and community involvement, and for engaging in ecological literacy. To this day, the Grandview/Uuqinak'uuh /Uuqinak'uuh garden provides an elusive bridge between practice and concept. Together with support from current principal Gloria Raphael, the Vancouver School District, and Coastal Health, garden coordinator Brent Mansfield will work with students, teachers, facility staff and community to continue to support the cultivation and mindful principles of local food as it relates to learning. As a garden that brings a strong cultural and historical legacy and valuable public support and funding Grandview/Uuqinak'uuh shows how an urban school garden can provide a leading example of local, grass roots sustainability.

Student Narratives from the Grandview/Uuqinak'uuh Garden

> The garden did not simply expand student vocabulary, but actually helped them find the words in the landscape. The students expression of space was entirely their own, as they kept journals and wrote poetry. They invited the poet John Oliver Simon to the garden and he told the students: "Explore! A poem can be anywhere in the Yard. Find your poem." (Moore & Wong 1997)

Today, as was the case when the garden first began, students at the school still face difficult socio-economic realities. In this case, student involvement in the garden began with simple awareness and connection as starting points. The students in this project visited the garden and described their encounters with nature and neighbourhood through oral stories and poems.

In the study, students were invited to write in their local environment/garden through a variety of writing activities that included free-writing on ecology-related topics including: bio-region, native plants, and endangered species. Most of the time we walked to the garden, to the schoolyard and surrounding neighborhood, noted the connections between each, and made lists on paper. The project recalled the words of Russ, the teacher at the Cesar Chavez school garden in San Francisco, where we would "just walk down.....and back."

About the Garden

> Don't look at our damaged petals and ask us why.
> Look at our garden and ask us how.
> (Kari Logue, "Roses and Concrete")

How did the collaborative, community vision of the original planners and designers unfold? How is the garden used today? In this garden, the physical space was expertly designed and built with much community and scholarly input. Work in the garden has continued. The garden at Grandview/Uuqinak'uuh still reflects the goals of a sustainable community that includes balance, harmony and respect for the socio-economic and botanical context. At this garden, the parts indeed symbolize the whole as true patterns in relationship. The "rooms" of the garden created by Tracy Penner are still multiple-use sites for the community. The garden is not simply a school garden, or a large-scale farm operation but a blend of ideas and principles that reflect, and bring renewal to, the surrounding urban area.

The garden coordinator at the time of this study, Christine, described the approach at Grandview/Uuqinak'uuh as unique and effective. Her role as garden coordinator is still a major part of what makes this garden unique; few school gardens have the dedicated support of one professional, hired specifically to teach and develop garden activities. Christine is aware that the large urban garden was at one time a big empty field. She described one of the benefits of turning an empty gravel field into a garden: increased positive foot traffic in the evenings, during summer holidays and on weekends. Christine explained that the ten garden plots are always in high demand by students' families and other community members. Normally, there is a wait list of approximately 30 people.

Where certain subjects like Science and Biology are taught more often in the garden, she has increasingly seen other subjects, such as Language Arts and poetry, incorporated into garden learning. However, Christine noted that the true benefits of the garden are often rooted in smaller-scale actions and relate to student motivation:

> The really huge affect I have seen, in practice, is with smaller groups of students, especially groups with struggling learners. They speak more openly than they would in an office.

Christine preferred to move beyond a basic goal of working to "just get the gardening done." When she brings students to the garden she prefers to let the students take their time. As she stated: "I come out here with students and we dig holes, find worms."

Christine discussed the challenges of keeping the area pruned, weeded and clear. The local police have said they need to be able to drive around the school and have clear sight lines through the garden. Given these urban practicalities, many large, bushy plants in the butterfly garden needed to be clipped back. However, the overall, original design of the garden, with themes of smaller units or rooms, still work effectively and spark the imagination. While we spoke, students roamed through the garden's smaller sections, rounding corners, and playing hide and seek. It is obvious that this garden, with its location and requirements for engaging marginalized learners, provides an alternative space that allows flexibility and balance.

Christine explained her perspectives of the role of gardens in inner-city schools, with many of her experiences sounding similar to other garden educators. One common theme in an inner-city garden is the scarcity of space itself:

> Most kids here don't have home yards. Their interaction with plants is little. Also, there is little park space in the neighbourhood. We don't have much green space. Turning the space into a space that is *actually used* is pretty amazing. It's beautiful to have a garden like this around for the kids. It has an affect on the kids that a big open field wouldn't.

FIGURE 17. Student writes poetry under cherry blossoms. Grandview/Uuqinak'uuh Elementary School Garden, Vancouver, Canada.

We discussed the need, and difficulty, of having to prove the tangible educational value of school gardens when in many ways the benefits are subtle:

> It's the kind of thing you can't measure....but it's also hard to deny. Being surrounded by that much beauty...you think how can that *not* have a calming affect on these kids, in particular, who have a lot of other things on their minds.

One of the other common, practical realities similar to many urban school gardens is the idea that already overworked teachers must involve themselves in "yet another project." However, the pressures on school/student measurement and performance standards are often antithetical to inviting the true potential value of learning in a garden. Christine supported the idea that teaching in a garden often begins with recognizing the simplicity of the space itself, with activities that do not necessarily require "special training" or advanced degrees in Biology or Horticulture. As Christine commented: "I mean, you can't mess it up that much. It's neat to give kids a handful of seeds and have them run around and plant them wherever. I don't even know where they plant them. I don't need to know!"

As the rain poured on yet another day in Vancouver's Grandview/Uuqinak'uuh garden, it seemed obvious that a garden rarely, and wonderfully, conforms to human notions of it. An urban school garden constantly tests teacher and student ability to allow natural processes into the everyday learning process. The inclusion of community, of diverse alternatives as focal points in an urban garden, the way a garden allows all students to learn and grow, the tenacity of a garden in challenging prescribed methods, are still the living principles at the Grandview/Uuqinak'uuh garden.

The original goals for the garden, current practices, dreams, and the dedication of the whole school community provide a backdrop for the students' responses as they visited the garden.

Student Reflections

At first the students wrote poems that named small, concrete details in their garden, which had previously gone unnoticed. From there, they made links to their neighbourhoods, and to their lives, using symbols from the natural environment. Their teacher, Sharon Vieira, described the transformation of the students as they began to write in the garden as "exceptional, dramatic, life-changing, huge." The main changes Vieira observed in these students, specifically in the context of this urban school, were related to concepts of "pride and freedom." She noted that when they began writing, their classroom participation and academic work improved with the rise in self-esteem. The First Nations co-coordinator at the school stated that writing poetry in the garden was a "safe way of journaling," a way to speak, and "hide" in metaphor and symbol. Over time, natural elements provided symbols for the often hidden socioeconomic burdens carried by many of the students. As their teacher explained: "They have a lot going on in their lives outside of school to process....every day. It has to go somewhere."

An emerging theme of the student writing was a juxtaposition of light and dark processes in nature:

> *The Sky*
> I see the sun shine in front of the clouds in the sky.
> Sitting out here in the cold, it fees like I am in the North Pole.
> This is great.
> The sky is gray and as big as ever.
> It was cold outside.
> I just had the goose bumps three times
> that's why I am called "Nature Cold."
> —Nature Cold, Grade 6

Permission to write openly about their own local environment, the garden/schoolyard situated in their neighborhood, meant their place, and their awareness of that place, became relevant, acceptable forms of academic discourse. Being tuned in to the context outside the classroom, and of the biodiversity found there, offered an important starting point for learners whose environments are typically expressed negatively, or ignored, by local media and school rankings.

> *A Poem to Rain*
> The rain sounds so quiet and has a low humming noise.
> I also hear a police siren in the background faintly.
> The rain makes trees look like they are dripping.
> When the rain hits puddles it looks like it's making bubbles.
> —Mr. Chris "A," Grade 7

In the inner-city classroom, students were delighted to hear the words that surrounded them, East Side street names like *Hastings*, *Rupert*, *Cassiar*, included in classroom discussion. The familiarity of the words always made the students laugh in recognition, perhaps because they had never before heard the names mentioned in the context of the curriculum.

When the discussion of bioregions lead to a discussion of bodies of water, the students talked about what they knew: Trout Lake, a small lake located in a nearby city park. I then asked the students to write a poem called "The Body of Water I Know." Many of the students focused on an even more familiar body of water: the puddles outside their homes, on their sidewalks, in their street gutters and school garden puddles. Instead of naming the larger local bodies of water (including the nearby Pacific Ocean and the Fraser River) the students looked into their own puddles, a largely unnoticed body of water right outside the classroom door. As they wrote in the garden, they constantly moved from local to universal ecological awareness with a sense of creativity, freedom and agility.

Body of Water
The bodies of water I know are the puddles.
It looks like mini lakes.
It sounds like a rain drop, I step on it.
It smells like mud.
The body of water I know is a puddle
that looks like the three great lakes of Ontario.
—Rosetta, Grade 6

Body of Water
The body of water I know is the Gulf of Mexico.
Looks like a giant bowl of water with a large piece missing.
Sounds like chaos and smells like fish.
The body of water I know is the Gulf of Mexico
and it's smelly, chaotic, and big but it will always be
the Gulf of Mexico.
—Antonio Montana, Grade 6

It was typical for the students to defy expectations, challenge all assignments, in order to make them their own. The students constantly surprised me with their creative interpretations of writing in the garden and their natural surroundings. I had never, until that day, thought of an urban puddle as a body of water, a concept that is so simple and true it seemed impossible to miss. Inclusion of their own, unique metaphors added to their body of writing on urban ecology which not only reflected their surroundings, but also their struggles.

One student tells a story about being chased by wolves at Mt. Currie. She said that one of the wolves forgave her friend for wanting to harm it. Outside, they stood under the cherry trees, looking up at the blossoms blowing in the breeze. Many of their narratives related to texture, to comparisons between the soft petals, and the hard cement. They made notes about all parts of the tree: the trunk, the blossoms. Being outside gave them the incentive they needed to work hard and to share their uniqueness in a safe, comfortable way.

We stood in the pouring rain, simply examining the natural space.

Rain
Oh rain you are so wet and
gross
When you visit I think I
can't do any thing
outside
I wish you would dry up
and go away.
You sound like....I really
don't know.
—AKA Shorty, Grade 7

AKA Shorty's poem, without an easy metaphor for rain, expresses the student's sense of ambiguity. Unconditioned to what a poem looks and sounds like, the student engaged in unsentimental, ecoliterate awareness, that projects an acute, close, honest awareness of her surroundings.

The descriptions of the garden, the surrounding neighborhood, the bleakness, the brightness, and the plainness of their worlds all carried meaning once they heard them read aloud. In fact, reading through all of the student poems, "contrast" was a common, ongoing theme:

Ode to Rain
Oh Rain you are so fast as you pat my head
and send a cold
echo through my body. When you visit I think
of a sad and hollow
day when no one cares. Sometimes I wish you
can just stop turning
the sky gray and stealing the sun. When I
watch you through my
window I only hear rapid taps and I see no
motion but falling rain,
And puddles. you sound like thumps over and
over.
—Antonio Montana, Grade 6

Ode to Rain
Oh rain you are so sweet and boring and loud too.
When you visit I think of little rocks hitting a vast drum.
I wish that you keep raining for plants and animals.
When I watch you outside I see little drops of rain and seeing the little circles on the pavement of the effects of rain.
I hear you sound like little rocks hitting a roof while sprinkling down the side
—Mister Roger, Grade 7

When reading the poems aloud, the students began to hear in their own work, and in the work of others, the connections between the students, their urban environment, and the organic contrasts in both. Closely observing the uncomplicated contrasts in the garden made students aware of the contrasts and changes within themselves. It was rainy one day, sunny the next, and both were accepted and acceptable, in nature, and in their own words. Difference, contrast and change, core characteristics of their everyday lives, all seemed to fit within the context of the garden. In many ways, the students developed a sense of wonder at just being outside, writing.

The students crowded under the massive, pink flowering tree outside the front of the school and studied the tree in quiet wonder.

This Fine Wednesday March 28 2007
Here we are with a pencil and paper
We went outside to the garden to discover
everything in the garden
At first it was calm now its loud

It felt cold but it warmed up
People were whispering and talking
I saw everything in the garden wondering
who, where, why, when and how it was made I
heard everything and still wonder.
—AKA Shorty, Grade 7

This Fine Day
We went to the garden to discover the beauty of nature,
At first it was quiet then it felt windy
People were quiet and still
I saw birds flying above the sky roaming around
—Bob, Grade 6

AKA Shorty's line, "I saw everything in the garden wondering," juxtaposes the student's sense of small and large, local and universal understanding. In Bob's poem, the student starts out on the ground, going to the garden, and within a few lines, "the birds are flying above the sky roaming . . ." The students uniquely describe both their awareness, and self-awareness, through the quiet in the garden, an attitude they did not often reveal inside the classroom. In fact, they were often reluctant and shy at showing any kind of subjective, individual, or emotional reactions inside.

Rainy Days
This cold rainy day full of cold suffering
so cold that no on can say a word . . .
The long silence hurts me from the inside out.
The everlasting raindrops falling on the unforgiving concrete.
A true way to show our feeling of the cold rainy days.
I see the water hitting trees so beautiful.
It reminds me of a waterfall violently hitting
the big hard undeserving rocks below. . . .
—Richelyn, Grade 7

Some of the most powerful poems from the students as they related to the natural world were about animals, and especially animals with symbolic importance. The students expressed their fears about extinction. When I asked the student to write "an ode to an animal" they cared about, most students wrote about wolves. The students talked about their fears as young urban citizens who care about their natural surroundings, with subtle recognition of the planet's inability to continue to provide habitat to wild species.

Ode to Wolf
Oh Wolf
You are so fast
And you're never last
You are so alive
I hope you survive
—Liamnardo Davinchi, Grade 6

Ode to Wolf
Oh wolf you are wise and swift
You are old and great
I hope you don't die out and stay alive
—Mister Roger, Grade 7

Ode to Animal
Oh thunderbird
You are beautiful as the moon
And you are stronger than steel
And a free spirit.
I hope that you get lots of room
For the young.
—Carnet, Grade 6

The poems also describe a sense of resilience found in their daily, urban ecologies. As they name it in their own words, they also re-imagine those ecologies. The student responses sparkle because the students found unique words for their own landscapes.

This Fine Day
On this fine day I see plants most dead
but some lively…I see birds traveling
and chirping. Also dirt as quiet as my lips are now
with a car beeping like a microwave when its done.
I see a skytrain almost Quiet as it passes by.
—Jessica Barrot, Grade 6

One day I asked the students to describe their classroom as an ecosystem. They responded:

Living Awesome Things
The parts of our classroom eco-system are the plants
and the people that surround us.
The moods of our classroom eco-system are busy and exciting.
I am part of the classroom eco-system
because I help and respect others.
Our classroom is a world within a world
because we listen to each others feelings.
—Ann, Grade 6

Another day in the garden the students responded to a question: What is green?

What is Green?
Green is jell-o.
Green is frogs.
Green is moldy logs from green bogs.
—Dark Soul

For these students, the garden provided a reference point for their deeper awareness of nature, in a uniquely urban Canadian context, where symbols of culture and deeper beliefs around the values of nature meet the challenges of the practical work of sustainability and environmental design. In the inner city, student poetry is another pathway to the garden, and to learning.

Lessons from Grandview/Uuqinak'uuh 's Urban School Garden

Qualitative thoughts and feelings cannot be ignored—we need words for them.
(Rowe 2006, 55)

As an ecologist and advocate of the imagination in learning, Edith Cobb said that "creative expression is rooted in a child's relationship with the complexity…of the natural world." (Cited in Evergreen 2000, 3) Cobb believed that immediate, experiential learning in the natural world also taught students about themselves. In such a view, nature is unmediated, uninterrupted by teachers and the student is allowed to encounter and engage. Such an idea formed the basis of the project at Grandview/Uuqinak'uuh. Permitting symbolic interpretation of local ecology invited students into the enchantment of their local urban surroundings.

While visiting Grandview/Uuqinak'uuh many people told me: "Things are different here." I soon learned that the story at Grandview/Uuqinak'uuh *is* different, in a positive way, and thus requires a different telling. The garden began with dedicated, innovative design, exemplary vision, solid funding and extensive planning. From the outset, the surrounding community, including parents, city parks and school board, scholars, and grass roots organizations were involved and lent strong support. The site was established with a vision that invited further involvement of pedagogical/community ideals that still continues. Such a design vision lead the way for diverse learning and research practices, for future story-telling and unique expression. One of the reasons why the garden here is unique is in fact related to its many unique challenges. As a community used to extremes in circumstance, the Grandview/Uuqinak'uuh garden represents a strong symbol of the empowerment, hope and potential that such spaces can bring to urban life.

At Grandview/Uuqinak'uuh, the practical work of storytelling took place in the garden, through poetry. In this community, difference and diversity is, in so many ways, the norm. The school has several unique challenges related to student socioeconomic status; the teachers and administrators are dedicated to helping student face these challenges, as a part of their holistic teaching practices. The principal at the time of this study, Liz Kloepper, explained that the staff measured student learning with "a wide angle lens," reviewing how far a student had traveled in a year. In such a setting, it is obvious that the students thought and learned in non-standard ways and, as a non-standard learning site, the garden has become a symbol of difference.

Contemplation of the local landscape led students to deeper understanding and engages them with their own learning process. As an ideal classroom, the garden seemed to justify, rather than intrude on, their awareness. The students at Grandview/Uuqinak'uuh very much interpreted the garden on their own terms. Whether considering a tree or a puddle, they found meaning in the nature of local neighbourhood. The culture and meaning of the space began with the students, who found ways to connect with their local surroundings.

Several bridges of understanding form a link between the founders of the garden and the way it is in use today. In one example, while the students had asked for a water feature in the original design, their poems ended up describing a different kind of body of water: the mud puddles. (While the Pacific Ocean, accessed on the other side of the city, was a world away) In another bridge to meaning, the students described how they wanted "places to hide" in the garden, which was provided by the garden "rooms" in the original design. In a similar way, the First Nations co-coordinator explained how the students used the poems from the garden to hide in metaphor, to safely express all they needed to express. Their sense of difference, in a space of difference (a garden), only made their work more powerful.

For the students at Grandview/Uuqinak'uuh, the garden and their expression of local environment are places to begin. The writing from this site shows how nurturing a plain awareness of the earth is possible in a way that nurtures learning in the inner city. Such awareness moves beyond institutional notions of sustainability to the potential for relevant discourse in marginalized, urban ecologies.

FIGURE 18. Classroom at Grandview/Uuqinak'uuh Elementary School, Vancouver, Canada.

· 7 ·

THE URBAN SCHOOL GARDEN

> The large is small in terms of the range of people's alternatives. The small is large where unleashing people's energies are concerned. (Shiva 2005, 183)

In light of contemporary environmental problems, the urban school garden brings limitless firsthand practice and knowledge of shared land use. As a site of learning, an urban garden reflects multiple purposes and concepts and becomes a community, a cultural artifact, a bridge between learner and land, and a place to incorporate new ecology-based curriculum. Gardens in urban areas involve teaching and learning with heart and vision while challenging learners to think about efficient use of natural resources.

The case studies illustrate how an urban garden begins with recognition of a local environment often made invisible. Every garden in the study describes a revolution in teaching and learning around the benefits of growing local food and simple observation of the natural world. An environmentally focused model of learning places equality, stewardship of nature, and honoring the environment itself, first. Today's themes are practical, include ecological and social justice and are often aimed at helping students, schools, neighborhoods and the larger planet, sustain itself. The goals in the modern urban school garden involve a re-prioritization of values where subtle concepts such as awareness, observation and connection are included and valued. As a student from Seattle said: "We are just talking right now. But it's still true." The urban school garden is a site that returns grounded ideals to teaching and research.

History describes immense institutional gardens on factory lawns where student labor was often used to benefit national economic visions. Over one hundred years ago, urban school reform aided training in urban communities to revitalize the economy of de-populated rural areas. In 1904 the school garden movement was called "the forerunner of a movement of deep significance and broad scope" (Klein 1904, 1). Food, and the success or failure of agriculture and farming, was widely recognized as a cornerstone of the economy. A school garden movement of "deep significance" in a time of renewed economic interest in agriculture very much reflects today's context. One major difference in the current reform is that grass roots community is central to the process.

FIGURE 19. Students at Kimbark Elementary, San Bernardino, weed garden beds.

An urban garden thrives in diversity and in retaining links to local surroundings. In all of the cases in this study the gardens were closely linked to the physical and social structures and to the goals and needs of the neighbourhood and city. The fascination of an urban garden as a site of study is that it is a place that emerges from its local surroundings yet offers a completely new experience of land-based learning. A garden represents an altered space, one of care and stewardship, that is also very much rooted in local community. As a place *within place* an urban school garden provides a visible example of how contextualized, local learning invites both difference and inclusion.

A garden's meaning (as practice, as philosophy, as a way of life) is itself a unifying force, filled with paradox. Gardens allow an approach to learning where teacher control over knowledge and individual power evolves into contextual, shared meaning. Such a new perspective also requires a renewed focus on human ideals in the midst of institutional teaching and learning: trust, responsibility, care, gratitude and heart.

Summary of Overall Findings

> ...Y me hice maestro,
> que es hacerme creador.
>
> ---
>
> ...and I became a teacher,
> in other words a creator.
> (Jose Marti)

All of the gardens in this study began with little or no budget. However, if thoughtful interaction between humans and land indicates a community's ability to sustain itself, during this study I observed teachers and students caring intently for every inch of garden in their stewardship. All of the teachers and students understood the connection between land, learning and the power of small-scale action.

While all of the teachers in the study highly value school gardens, no one believed that gardens should become a mandatory or enforced curriculum standard. As sites that bring freedom, innovation and enhanced learning, all of the teachers felt that inclusion of school gardens into regular classroom work should arrive by choice. The teachers described how firsthand experience, for both students and teachers, was the way to discovering how such sites can potentially transform teaching and learning. Such awareness must be invited and arrives gradually through experience.

The teachers spoke about the value of their gardens in a variety of ways, for the way they inspired intrinsic value, practical value, motivation, community, social skills and a sense of belonging. In each garden, the site represented different ideas, in different ways. None of the teachers had any desire to re-create paradise. All viewed their gardens as valuable, hard-working spaces that enhanced teaching and learning in both planned and unplanned ways.

Each garden held a unique focus and contribution. Kimbark in San Bernardino contained a variety of gardens, scattered across the campus; their program has included field trip programs for students across the city. El Dorado in San Francisco has long cultivated connections to the Greenway, to the local park, and to the neighbourhood. Programs at Maple in Seattle arise out of a strong tradition of environmental learning and practice in the Northwest while future plans for their garden include creating an ecological learning site. Grandview/Uuqinak'uuh is closely linked to the local First Nations community, to local food connections and distinct, school district programming that also serves the neighbourhood. In all cases, supportive educational leaders nurture the innovative ideas of individuals that ultimately promote and advance those innovations in the community.

The narratives of this case study show multiple perspectives around the cultivation of land for the purposes of learning. From the farmer in South Central Los Angeles ("I never thought that growing food would make you so political") to the teachers describing the value of a garden in the city, all of the stories assert the basic human right to survival within collective cultivation and participation in public space. The teachers told me their priorities in teaching in the garden: community, stewardship, involvement, empowerment. All spoke about the need for teacher flexibility in teaching in a garden. They told me: "the sidetracks are important."

While landscape design was a valuable aspect to the success of their gardens, physical design of space was holistically integrated into the planning, pedagogy, practical concerns, and community goals for each garden. At El Dorado, the new serpentine shape of the garden is an intentionally planned addition to the former site that allows several students to take part in the small space at the same time. At Grandview, the holistic landscape design provides a leading example of culture-land-community integration that still influences how the garden is used today. At Maple, the garden includes elements of archways, small gathering places, and is located beside the garden teacher's classroom, with room to expand into an adjacent field. The Kimbark garden, the oldest garden in the study, was established at a distant point from the school, on a hilltop. The

Kimbark garden now provides a transitional space where students walk between school and garden.

All of the teachers described the value of grass roots involvement. They have plans for their gardens that will require strong support for their existing structure and organization so they can focus on what is truly important: connecting students to learning, local food, land, culture, history and to their own survival. All understood both the realistic and the subtle benefits of the garden. They know that the skills learned in the garden may not guarantee jobs to their students in future yet all teachers described how the students learn skills that improve their lives, right now. All saw the gardens as a place for urban students to begin.

The reality for hard working teachers often means budget cuts, layoffs and even the elimination of entire programs. The teachers and educators in this study all worked hard to create enriched learning environments in the midst of severe cutbacks to education. Yet, all of the teachers are unwavering in their support for quality education and for community networks of learning. Most of the teachers developed their gardens on shoe string budgets; school and community support is key to their survival. When their work appeals to the goals and values of their communities, their projects thrive.

The experience of the teachers is reflected in current school garden research where the principles of curriculum design begin with ecoliterate teaching activities that include hands-on involvement. (Capra 2005). Eco-centred activities require community-based, process-oriented teaching and learning. Such ideas also coincide with "the slow school" movement where "Students have time not just to memorize, but also to understand." (Holt 2005, 60) Even as a teacher facilitates the flow of activities, an outdoor classroom changes the role of the leader because learning tends toward collective decision making. While various models promote school gardening, the teacher's role in learning in a school garden involves guiding and facilitating interdisciplinary knowledge with an emphasis on practice.

Garden pedagogy always includes the changing processes of the gardens themselves. Urban garden educators often teach in small settings within inner cities; they also teach with imagination and willingness to learn beyond the parameters of prescribed boundaries. In turn, teachers inspire the same sense of determination and mindfulness in their students. As one educator said, building a garden was like "planting a seed," where students could build a foundation—for life.

Student Responses

> We cannot dig, plant, trim, water or harvest with detached passivity. Gardening can hardly be done without getting hands dirty, and, in most cases, getting earth under fingernails and blisters on palms. This is what many gardeners report as the essence of their actions. (Francis & Hester 1990, 6)

In cities, students connect with gardens in ways that symbolize larger connections to the natural environment. In 1901 a student in a garden made the following observation in a notebook:

> Rabbit, squirrel, and gopher.
> Cow, sheep, and goat.
> Hen, duck.
> Frog and tadpoles.
> Bees, caterpillars, and butterflies.
> (Klein 1904, 96)

Now, a student writes:

> You see
> how it kind of shines
> if it's watered
> and if its wet you see
> how the leaves shine.
> (Maple Elementary, Seattle)

Close and careful observation is found in both narratives. In the current case studies students took note of their surroundings and also described their connection to place and to learning *beyond* the garden. The students talked about beans, carrots, soil, water and their connection to land. They wrote about gophers. They talked about what they had seen and learned and applied that information to their work inside the classroom. Their lyrical responses describe their sense of pride, dignity, and respect for local environment. An examination of student responses from the schools reveal certain common themes.

The students described strong understanding of local conditions for growing food. They focused on their sense of pride in growing vegetables themselves: "You actually get to see it: How to plant a seed." (Kimbark); "We can grow the food and cook it. Hey!"(El Dorado). Students also showed understanding of farm practices linked to responsible stewardship: "Don't pick unripe strawberries don't step on the fruit."; "I learned how to compost old weeds. Don't kill on purpose." (Kimbark).

A common thread in the writing is the connection to their urban surroundings and sometimes to the danger they feel in daily life. As one student wrote, in the garden "You can be un-scared of anything" (Maple). When asked to write an ode to nature, instead, they chose to write about their fears of nature's extinction: "Oh wolf you are wise and swift. You are old and great. I hope you don't die out and stay alive." (Grandview).

Students also expressed admiration for the powerful ways that plants and animals survive. Their understanding of the principle of cause and effect link to values of responsible land stewardship: "Yeah! Without the garden we'll die."; "Fruits have seeds. Fruits come from flowers (so don't pick the flowers)."(El Dorado). "We work hard, we eat fruit."; " "You are so alive. I hope you survive." (Grandview).

On the question of learning "inside or outside" many students responded by describing sensory awareness: "I work harder outside. Inside a lot of things on a paper."; (Kimbark). "I can see things in a garden I cannot see in a classroom."(El Dorado). Learning inside includes reference to fixed structures ("inside: bricks, and math") while outside students describe their movement into the world: "Here, you walk around" (Maple). "It sounds like a rain drop, I step on it. It smells like mud."; "Sounds like chaos." (Grandview)

The garden also brings knowledge of community, empathy and safety: "Sharing is the key." (Kimbark); "Even if you have a store with no corn you can always ask your friend: 'some corn seed please.'"(Maple). The students also expressed a complex understanding of space. By moving inside and outside, students learned to transfer their learning; and, they learned to transfer values between garden and classroom. "Big! Big! It's really Big" (Kimbark). "The environment, it grows into what is around it." (Maple). While students are able to apply and transfer practical knowledge in the garden they expressed awareness of the connection between theory and practice: "Right now we're just speaking of stuff. (But it is still true)"; "If you see everything here you see everything everywhere." (Maple)

Another strong theme was student sense of pride and empowerment in their own urban surroundings: "We have the best garden teachers in the world" (Kimbark). "Every school should have one. (Yeah!)"(El Dorado). "World: we have plants. We have a garden. And we help the environments."; "Out here....I got the uniqueness." (Maple) "I saw everything in the garden wondering who, where, why, when and how it was made. I heard everything and still wonder." (Grandview)

Learning the principles of cause and effect in the garden, students learn to admire organic processes. One of the most prominent lessons of the garden relates to the student connection to the earth. They spoke about the impact of their light ecological footprint. "Outside we help the flowers breathe"; It brings in the world.…I like the clouds in the garden" (Kimbark). "Compost. Live it"; "Outside! The garden has all the self with nature" (El Dorado). "The life of the plant? It's all full of life!" (Maple); "Our classroom is a world within a world because we listen to each others feelings."(Grandview)

A sample of student responses shows how firsthand experience of gardens brings practical and complex understanding to learning on a daily basis. During the interviews, students spontaneously described their practical garden knowledge as being linked to respect for life, to their own surroundings and to a sense of pride in their own space. In their school gardens students were able to tell their own stories, in their own ways. The students' responses show how they cannot dig worms without valuing those worms. They cannot describe how they grew carrots without also describing pride in growing carrots.

Summary

- **School gardens help students connect their learning to the world while also teaching skills of adaptation.** Observations in this case study mirror the latest findings from the Royal Horticultural Society (2010) which emphasize how gardens help students: overcome fears, transfer their knowledge, develop patience, and adapt to changing learning conditions.
- **A history of school gardens describes the practice and pedagogy that helped shape today's perspectives on gardens.** Common themes that influenced school garden movements between the turn of the 20th century and into the 21st century include: community, economy, food, and a need to increase access to nature in the daily lives of urban children.
- **The main differences between historical and current perspectives on school gardens also relates to themes of community, economy and food.** The historical focus on economy related to large-scale, national concerns. After a century of declining interest in a food economy, perspectives today centre around small-scale, local, sustainable economies with renewed student and teacher engagement in local farm traditions.
- **Gardens are rooted in local community.** As a community-based initiative, learning in school gardens has shifted to include informal, student and local knowledge.
- **A garden promotes reflective learning, as process.** Perspectives of front-line educators consistently describe how a garden promotes experiential learning in process and makes the process visible.

- **Research in gardens requires front-line work alongside garden educators and students.** The work of gardens nurtures processes of conceptual work, grounded knowledge and physical labour; garden research requires the inclusion of all such elements.
- **Gardens change research.** A garden has its own knowledge, unique to each region. The so-called inner-city school stereotype prevalent in much Education research is challenged by ecology-based learning. Urban school gardens today reflect enthusiasm, creativity, integrity, excitement, engagement, health, love, wonder and hope.
- **Place changes learning.** As a place *within place* a garden transforms community. When students lack place, and would not otherwise have access to nature, a garden has a large influence on students' lives and learning. In this way, a garden becomes a symbol of hope and transformation while remaining part of that community.
- **A shared community garden revitalizes the commons by revitalizing commons ideals.** A garden in an urban area supports the basic elements of survival that are sometimes forgotten in the politics of land, food and place. A shared garden revitalizes concepts of community, care, trade, farming, and de-centralization of land ownership.
- **A garden addresses local (and global) urban needs.** In urban settings, in areas where food security is threatened for a variety of reasons, a garden represents a balance where students' most urgent needs are addressed within the context of learning.
- **Gardens promote difference.** The students in this study were all eager to express their gardens in a certain way—either through song, poetry, art or story—and these ways must be heard and honoured in order to truly understand the transformative pedagogy of a garden.
- **Garden educators work with students starting with "where they already are."** While urban problems are large, complex, and typically ignored in mainstream curriculum, a garden gives students a sense of personal power. Nature's lessons, learned at the ground level, are open to all.
- **Some historical ideas mirror current ideals in garden-based teaching; some ideas differ.** Many progressive ideas of the last century are still considered alternative. Today, many educators similarly seek alternative connections and meaning.
- **Gardens move learning outward and transform teaching and learning.** Many of the teachers described how new ideas largely emerge from taking note of what is already taking place in the garden; a learner's role thus becomes focused on experience and observation.
- **To understand the small garden is to understand the cultural and pedagogical significance and potential of all gardens.** While the examples of larger gardens describe one kind of evolution of the movement, smaller gardens contain the seeds, the origins, the original desires and ongoing ground-level experimentation of the cultural movement as a whole. Future research, lies in widespread, small-scale, community innovations.

- **The solution can be small.** A small urban school garden can produce far-reaching effects. It can lead to knowledge of practical techniques for growing food, to understanding of local ecology. A garden can invite self-expression when that is what is needed most. When the learner and community is surrounded by challenges that seem insurmountable, small scale responses become a means of survival. Or, as one of the students at the Grandview/Uuqinak'uuh garden said:

The gravel field
as big as my heart
and yet I can't find it.
—Jellybean, Grade 7

New Perspectives Needed

> I have learned that any sense of *who* a person is, and the standpoint he or she occupies, has a distinctive bearing on the kind of stories they tell. *Who* cannot be separated from *where*. The better we attend to our locationality, the better we are able to understand what is at stake in our lives. (Donald 2004, 1)

In the past, research has typically rewarded science and ideas that promote capitalist economic growth without any view toward the consequences or ecological or social impact of scientific innovation. Schooling, on the other hand, often turns to ever more standards-based solutions to solve problems related to lack of student engagement that breed in such a system. As for the gardens themselves, arguments that question the value of urban school gardens tend to dismiss the extensive qualitative evidence that supports school gardens while ignoring both the food crisis and deeper learning significance. In some cases the notion of linking farm skills to schooling is dismissed altogether. One article in *Atlantic*, "Cultivating Failure: How school gardens are cheating our most vulnerable students" (Flanagan 2010) derided the school garden movement for its involvement in farming, described in the article as "stoop labour." Typical criticism of the movement includes the assumption that school gardens will replace all other learning traditions. A related argument against school gardens involves the perceived inability to prove, beyond a doubt, that school gardens help students achieve standardized outcomes.

In reality, school gardens have, for some time, served multiple, civic, educational, economic and vocational purposes. Now, these purposes are once again seen as mutually beneficial: food farming brings pedagogical understanding; students learn to learn from their natural environment; students learn to honour nature by working with the land. Over the course of one hundred years, popular focus has also shifted from "saving" urban youth to saving the planet. Instead of an entrenched, binary focus on either economic concerns, or moral ideologies, the rise in urban agriculture and sustainable practices closer to the turn of the 21st century has shifted to support and nurture the practical value of sustaining the planet. Such interest is widespread in popular society, in community organization, and amongst youth themselves.

Historically, curriculum has long reflected the community that surrounds it. Outcomes and standards emerge from economic necessity, community ideals, and current ideological beliefs. Yet, the thinking that values decontextualized learning, standardization, test scores and a race for individual or political dominance tends to ignore local, ground-level, shared, tangible learning experiences and the first-hand voices of students and educators that support such experience. Traditional First Nations elders and some environmental educators teach that the problem facing the planet is a lack of respect for the natural world. No matter how well intentioned, a garden-based curriculum based on standards and outcomes that leave little room for nature's lessons is a replication of previous teaching and learning processes. Without a deep, long-term transformation in teaching, learning and pedagogy, as was the case in previous garden reform movements, once the visible "crisis" is perceived to be over, the garden movement will also pass. Now more than ever, due to environmental necessity, curriculum and schooling needs to be re-framed and re-oriented toward earth-centred practice and thought that will ultimately help address changes to the planet while also helping humanity learn to understand and honour the environment.

One of the main findings of this study was the role of community in supporting innovations in schooling. While the urban school garden emerges from the history, pedagogy and story of such spaces, the survival of such spaces is increasingly rooted in the surroundings themselves. The focus on curriculum integration, school board and civic support is valuable in helping students potentially access environmental education. However, as history shows, the involvement, and sometimes passing interest of such entities do not always represent the values needed to sustain the movement. What in fact keeps gardens alive, and learning alive in gardens, over time, are community ideals which only inspire more idealism, more involvement, and more gardens. Because the lasting value of a garden is usually found in non-standard and informal ways it is the smaller stories of a garden that require care and attention as guiding research principles.

The larger narratives around how we work within and describe the planet

FIGURE 20. Students at Kimbark Elementary School, San Bernardino, California. Tracy L. Penner, Landscape Architect.

itself are changing. Previous ideas on large-scale solutions within the same centralized power structures, do not, in fact, fit the micro-scale and diversity of nature's systems. (Martusewicz 2006) The pedagogical value of urban school gardens, then, lies in understanding and recording how local, collective action fits into the bigger picture of social reform, environmental design and sustainable learning. Students define and re-define their own learning through shared labor and an ethic of care based on their own unique experiences and urban ecosystems. In this way, the case studies in this book help provide examples for the future.

Cyrene St. Amant, a former classroom teacher in Watts, South Los Angeles (1990–1999) and now a teacher at the UCLA laboratory school, describes how urban school gardens directly benefit the surrounding community:

> For all city kids, just making that connection to *where food actually comes from* is huge....For inner-city kids, there is an equity issue as well. There are more farmer's markets in low-income communities now than there were when I first started working with school gardens. Although it has begun to change somewhat, traditionally, inner city populations have not had access to fresh, quality produce

An urban garden community connects people to the earth and helps them feed themselves; along with the current, widespread popularity and civic support for community garden and greening projects it is vital that such projects remain in the hands of their own communities. In Los Angeles, the South Central Farm was a proven way for a community to feed itself, yet it still faced the problem of land development and relocation.

The future of urban gardens will continue with roots in ecological and social justice, exemplified in Detroit and now rising in other cities characterized by strong grass roots vision and local action. Such projects mark a clear evolution from turn of the 20th century urban school gardens because they are built by and for communities and increasingly include partnership with academic research networks. The inclusion of first-person narratives in environmental education research help illustrate how grass roots projects must remain rooted in local community. Director of the Detroit Black Community Food Security Network, Malik Yakini, points out:

> We are interested in food justice, not just food access and increasing food availability. Food justice is about upholding dignity and helps to empower the community. Profit from food grown and sold in the community should re-circulate in the community, which in turn helps to empower the community. (cited in Kanter)

In Detroit, the citizens themselves have created a small-scale, model green economy based on ecological design principles and organic farming. These smaller farms do not generate profit and use multiple small plots, provide local food and urban farming skills while remaining true to the ideals of social and ecological justice. More recently, however, a large-scale, corporate development plan proposes a centralized farm development that would displace Detroit residents and reduce the democratic control and access they now have to small, community and school farms. The tension between Detroit's existing, small-scale farms and proposed agri-business illustrates the difference between the values and principles that surround urban public land use. The Detroit model also raises new questions, not simply about land efficiency and economic profit, but about social and ecological ethics for rebuilding a vacant inner city.

Can both models exist? Malik Yakini, of Detroit Black Community Food

Security Network points out that plans for the proposed urban agri-farm in fact work against the very principles of urban, community organic farming:

> One, there is little to no community engagement: farms in people's neighborhoods impact people's lives. These people have a right to be informed and have a voice. Two, [the project] does not plan to use organic methods and this is troubling…(cited in Kanter 2009)

Yakini also notes how important it is for the Detroit community to be able to maintain local control over their own food. In working alongside students and residents of inner-city Detroit on school and community greening and garden projects, researchers at the Southeast Michigan Stewardship Coalition have observed the results of urban agricultural programmes in Detroit. They describe the varied benefits of the garden projects and the new economies that spontaneously flourish around it:

> A network of people across the city were working to take back their neighbourhoods by growing gardens that fed their families, creating public murals, converting abandoned storefronts into studios, shops and educational facilities. They worked in small independent groups to revitalize their commons often through barter relationships, or giving work for work, and all the while educating each other in the process. (Martusewicz 2008, 20).

When students, youth and communities engage in publicly held urban gardens, they learn to produce food and feed themselves; they also connect to the earth in ways they may not have previously experienced or have been prevented from experiencing. The educators, volunteers, and administrators in this study understood the importance of equal access to the natural environment; they understood that survival in a city depends on access to food. In all cases, gardens were viewed as more than a simple teaching or landscaping tool and necessary to urban survival itself. Their knowledge and hard work in small spaces, with few resources, and a local model of decision making reflects a green economy and all that is now taking place, at a larger scale, in cities across the country. Over one hundred years of urban school gardening show how these spaces always relied on human support from the community and the school for their survival. The future of the urban school/community garden movement, more than anything, depends on local, grass roots action and must retain that same control and spirit. The idea of creating large-scale corporate models of sustainability may, in a capitalist economy, seem like a natural progression for the urban farm movement. However, what the gardens in this study help demonstrate, so powerfully, is that one small garden built within, by and for an urban community has a lasting value and learning potential, precisely because it is small and resides within local community. Such a model can be replicated over and over, in every small corner of every school across cities and nations.

Good Dirt

> In the future the garden will take on new oppositions, paradoxes, and muses; but it will still be profoundly sunshine, skill, love, struggle and good dirt. (Francis & Hester 2004, 19).

In the post-Victorian era, when society became more mechanized, the notion of simple observation in gardens and parks was considered passé and old-fashioned. Similarities exist today in school curriculum and planning, where the idea of *nature appreciation* in itself, is rarely validated. Some researchers believe that in the frantic pace of many schools the "Talk of slowing things down, dwelling over something and deepening our experience of it begins to sound vaguely quaint, antiquated, and simply unrealistic." (Jardine, Clifford and Friesen 2008, 12) Learning through close interaction with the land invites educators to engage in a wide range of new, progressive definitions around teaching and learning that include an "ecological interpretation of the basics" and a "collaborative intelligence" (Ibid., xiii) that involves an exchange between teachers, learners and landscape. The students/teacher narratives in this study reveal how the gardens themselves often seemed to exceed their expectations and inspire a change in consciousness with greater value placed on local, small-scale action and learning. Simply allowing gardens to grow, and students and community to help, transformed student and teacher consciousness.

What started as a study about urban school gardens became an exploration of cities and place and the potential to connect people with environment through a balance of teaching, learning, philosophy and research. Pedagogical research in urban school gardens is a relatively new field that involves a wide range of perspectives and interdisciplinary, holistic approaches. New environmental problems bring new goals and require new ways of measuring those goals. Such research begins with the idea that gardens create connections as the basis of life and learning. The gardens in this study represent a move forward in pedagogy as learning is linked to basic human interaction with the earth at the level where the earth is experienced, sensed, and shared.

Emerging pedagogical research in urban gardens will take on new shapes and will involve new collaborations, now being developed, that include the integration of formal and informal sectors, across school, university and community groups. The academic field will also be influenced and perhaps led by community organizations already involved in applying progressive knowledge to practice in local urban space. A school garden will thus become a focal point, a neutral ground, and an open space for ideas that foster and further academic/community collaboration.

In gardens, power is shared. And while the complex meaning of gardens is ever present, the garden always returns learning and life to the ground level, back to the soil and weeds. Practice and knowledge in school gardens includes ground-level experience and permits pedagogical ideals to flourish in a way that also brings a new consciousness. In the midst of the natural beauty of North America—a place that also includes massive concrete freeways, strip malls, and intense real estate development—the first challenge might be to simply invite students to the land.

This study began with questions about urban school gardens. However, learning in gardens in fact does not lend itself to easy definitions or to a formula for success. Meaning emerges in a garden's struggle to exist. When place itself becomes part of the lesson, learning takes a new direction. During this study I met individuals who developed, built and maintained gardens despite all odds. Always guiding their work was a way of viewing the world that was hopeful, passionate, and rooted in the land and learning. Their actions were intentional and filled with care. The story of urban school gardens is in fact a story of care and hope in solving both the practical dilemmas of urban food sourcing as well as addressing "the crisis in how we think" that helped to create environmental damage.

A garden is thus a garden, and more. A focus on the cultivation of food alone, at any cost, can easily overshadow the long-term, cultural and economic needs of a place. As seen in the current case of Detroit and, increasingly in other inner cities, to put forward solutions that largely serve the needs of industrial capitalism, ignores a community's own interpretation of the land as it helps them survive and make a living. Currently, urban gardens potentially bring new knowledge to a new, green economy that must remain rooted in widespread, small-scale work in a way that continues to serve local needs. From New York, to Detroit, to San Bernardino, to San Francisco, to Seattle, to Vancouver, and beyond, the local histories, cultures, pedagogies and practices within such places must be made visible.

> Through our Oral Traditions we know that the river or the mountain is our ancestor, and that beneath our waters, spirits live. It is precisely because the river is our ances-

> tor that we have an obligation to protect and guard the spirits that live within the waters.…(Carr 2008, 137)

Urban ecological literacy begins with schools and neighborhoods, with respect for what a community already knows. If the practical solutions are small and rooted in local communities, then environmental learning begins where the learner begins. As seen in this study, when stories about the urban landscape become part of everyday cultural learning in schools students are then able to re-interpret their local environment on their own terms. Students described their local ecologies from the ground level in a way that made a transformation seem possible. Ultimately, reclaiming a common space in an urban landscape, both in action and in idea, is a return to the land, to community-based knowledge, and a return to common belonging. As the natural environment continues to change, as the food crises in cities continue to provide new and urgent dilemmas, a change in both behaviour and thought is a practical requirement for survival.

> *World, I love this garden.*
> —Student, El Dorado Elementary School

APPENDIX

Selected Resources (National)

National Gardening Association
www.kidsgardening.com/

Includes information for teachers and parents about the benefits of school gardens for kids and youth. Includes project ideas and activities, classroom stories and more.

Youth Gardening—American Horticulture Association
www.ahs.org/youth_gardening/

American Community Gardening Association
communitygarden.org/

Community Garden Start-Up Guide: University of California Cooperative Extension
celosangeles.ucdavis.edu/garden/articles/startup_guide.html

Education for Sustainable Development Toolkit
www.esdtoolkit.org/

A manual for educators and communities on how local sustainability goals can be reached through collaboration and community partnerships.

Energy Hog
http://www.energyhog.org/adult/educators.htm

An educational site for grade 2–6 teachers and students that includes both teacher and student guides, activities, handouts and quizzes.

Botanical Society of America
www.botany.org/

The Center for Ecoliteracy
www.ecoliteracy.org

The Edible Schoolyard
www.edibleschoolyard.org/homepage

Center for Environmental Education
www.cee-ane.org

Sustainable Agriculture in Education
www.sagecenter.org

Catherine Ferguson Academy for Young Women in Detroit
http://growindetroit.filmmij.nl/school.php

Southeast Michigan Stewardship Coalition
http://www.glstewardship.org/About/RegionalHubs/SoutheastMichigan.aspx

The Detroit Black Community Food Security Network
http://detroitblackfoodsecurity.org/

DC Food for All
http://dcfoodforall.com/about/

Project FEED
http://web.pdx.edu/~feed/mission.htm

SPIN Farming
http://www.spinfarming.com/

Small Plot Intensive Farming explains inexpensive vegetable farming techniques specifically for land under an acre in size. Links to many North American small farmers who use the technique.

Canadian Wildlife Federation
http://www.cwf-fcf.org/
Gives information about wildlife, habitats and other wildlife issues. Articles are available on a range of environmental and wildlife issues. The site also provides 29 lesson plans for teachers on wildlife and environmental-related topics.

David Suzuki
http://www.davidsuzuki.org
Includes a wide variety of environmental information, practical steps for slowing climate change, and contests. Provides students with practical suggestions for environmental stewardship. Ten practical challenges and corresponding activities are presented to kids to conserve energy and be friends of the environment. Links to games, quizzes and environmental information are included as well. Sources for parents and teachers regarding environmental education can also be found on the site.

Evergreen
http://www.evergreen.ca/en/
An extensive Canadian environmental site with links to school garden projects.

Green Teacher Magazine
http://greenteacher.com/
A quarterly print publication for educators. Each issue includes: Ideas for rethinking education in light of environmental and global challenges, practical articles and ready-to-use activities for various age levels from 6 to 18, and resource reviews and evaluations of dozens of new books, kits, games and other green resources.

Seeds of Diversity
http://www.seeds.ca

Tree Canada
http://treecanada.ca/

Regional Resources

Southern California

Common Ground Garden Program/Gardening Angels
http://celosangeles.ucdavis.edu/Common_Ground_Garden_Program/

Los Angeles Common Ground is part of LA county's University of California Cooperative Extension, and trains master gardeners in low-income neighborhoods in LA, Lynwood, Compton, Inglewood and Long Beach. The group facilitates community gardens in schools and homes and set up a group called "Gardening Angels."

Farm to School Network
http://www.farmtoschool.org

In the mid-1990s this group was called "LA Harvest," a farm-to-school pilot project that began in LA (Santa Monica-Malibu USD) and Berkeley. Now, the National Farm to School Network is a collaborative project of the Center for Food & Justice (CFJ), a division of the Urban & Environmental Policy Institute at Occidental College and the Community Food Security Coalition (CFSC). Since the group started with a handful of schools in the mid-nineties, just over a decade later, the network now involves over 2,000 schools spanning 40 states. The goal of Farm to School is to bring healthy food from local farms to school children. The program teaches students about the path "from farm to fork," and actively promotes local produce in school meals to mitigate environmental impacts of transporting food long distances.

Food from the 'Hood
http://www.foodfromthehood.com/

Food from the 'Hood (FFTH) is the first student-managed natural food products company in the U.S. FFTH began in 1992 as a classroom project at Crenshaw High School in response to a Los Angeles uprising. The program teaches work-based skills training, academic tutoring, life-skills development and practical business experience. The students began to market their own line of salad dressing with help from local businesses. To date FFTH has awarded over $140,000 in college scholarships to student-managers.

Los Angeles Community Services Unlimited
http://www.csuinc.org/

LA Community Services Unlimited began in 1977 to develop community gardens in South Los Angeles. The program started at Normandie Avenue Elementary, where students worked with the community gardeners from the neighborhood to build one vegetable bed—the program has now expanded into an after-school program with fruit orchards. The goals of this group are ecological, social and community based. The mission of the group is to foster the creation of communities actively working to address the inequalities and systemic barriers that make sustainable communities and self-reliant lifestyles unattainable.

From the Ground Up
http://www.csuinc.org/fgu.html

From the Ground Up is a project of Community Services Unlimited partnered with Food Forestry International. The group developed three urban mini-farms and a produce market in South Los Angeles that provides food, employment and increased community access to fresh food.

South Central Farmers
http://www.southcentralfarmers.com/

When it started in 1992 South Central Farm was the largest urban farm in the United States. From an empty lot on city land the farm grew into a small-scale economy, covering a large city block in South Los Angeles where local, immigrant families or those with restricted incomes were able to grow food to feed their families. Since then, the farm has relocated and continues to support local urban families.

California Foundation for Agriculture in the Classroom
http://www.cfaitc.org/

Includes K-12 lesson plans

California School Garden Network
http://www.csgn.org/

Includes resource and publication links to help with your own school garden community project.

Northern California

SLUG (San Francisco League of Urban Gardeners)
http://www.grass-roots.org/usa/slug.shtml

SLUG was the original hub for urban community gardens in San Francisco. Now the group is a grassroots organization focused on education and employment opportunities around urban community gardening and ecological sustainability.

Garden for the Environment
http://www.gardenfortheenvironment.org/

Garden for the Environment's role in urban school gardens is around agriculture and education in a specifically urban context. The non-profit organization operates a one-acre urban demonstration garden and offers environmental education programs about organic gardening, urban compost systems and sustainable food systems. Since its founding in 1990 the garden has operated as a demonstration site for small-scale urban ecological food production, composting, organic gardening and low water-use landscaping.

The group also started the Victory Garden in the San Francisco Civic Center Plaza and a new Victory Garden located on the UC Berkeley campus, close to Memorial Glade.

Alemany Farm
http://www.alemanyfarm.org/

Alemany Farm is a 4.5-acre project in South San Francisco that grew out of St. Mary's Urban Farm. The farm is now run by the Alemany Resident Management Corporation, a non-profit organization dedicated to improving conditions in the Alemany Community, a 165-unit public housing development. The Farm is situated between a major freeway, busy Alemany Boulevard, and a housing project below Bernal Heights. The farm encourages local residents to volunteer, to train in the garden, and to take part in growing and selling the food.

Life Labs
http://www.lifelab.org/

Life Labs is a well-known, nationally acclaimed Center of Excellence program on the campus of UC Santa Cruz. The program started out as a small nature-based education program in 1978 at Green Acres Elementary school in

Santa Cruz, California. The project soon grew to involve twenty-five K-5 classes through a community and university partnership. At this site, children and teachers explore a variety of life, earth and physical sciences in what is now a two-acre living laboratory of flower and vegetable gardens. Life Lab is also well known for its research into school gardens as site of effective experiential learning.

Oakland Food Connection
EC Reems Academy of Technology and Arts (East Oakland)
http://www.foodcommunityculture.org/

The goals for the Oakland Food Connection program are agricultural, educational, nutritional, social and revolve around developing social justice and access to food for the local community. The rooftop garden at EC Reems Academy of Technology and Arts in East Oakland is a project of Oakland Food Connection. The site trains students in sustainable urban rooftop gardening, and the site involves direct access and participation in local, urban food sources.

SF Gro
http://www.sfgro.org/ua.htm

People's Grocery
http://www.peoplesgrocery.org/article.php?story=links

Northwest

P-Patch
http://www.seattle.gov/Neighborhoods/ppatch/

Seattle's P-Patch program is operated out of the City of Seattle Department of Neighborhoods in conjunction with the not-for-profit P-Patch Trust. The program currently provides sixty-eight gardens for residents throughout Seattle with a focus on community gardening, market gardening, youth gardening, and community food security in the City of Seattle.

Tilth
http://www.tilth.org/

Oregon Tilth is a nonprofit research and education membership organization dedicated to biologically sound and socially equitable agriculture. Oregon

Tilth offers educational events throughout the state of Oregon and provides organic certification services to organic growers, processors, and handlers internationally.

Seattle Tilth
http://www.seattletilth.org/

Seattle Tilth is a nationally recognized non-profit educational organization dedicated to inspiring and educating people to garden organically and conserve natural resources. The program offers classes, gardens, demonstration events and other support for local organic gardening. The mission of Seattle Tilth is to inspire and educate people to garden organically, conserve natural resources and support local food systems in order to cultivate a healthy urban environment and community.

Learning Gardens Laboratory, Portland, Oregon

The location of the garden in South East Portland serves students and families in the city's urban, lower-income neighborhood. The garden works with Portland Public Schools to serve adjacent Lane Middle School. The Learning Gardens Laboratory started in 2005 with a partnership with Portland State University, Portland Public Schools (PPS), and the City of Portland. The garden is part of a research project in Portland State University's Educational Leadership and Policy (ELP) Department in the Graduate School of Education.

Puget Sound School Garden's Collective
http://www.pugetsoundschoolgardens.org/

Seattle Youth Garden Works
http://www.sygw.org/home

Farm to School Oregon
http://www.farmtoschool.org/index.php

Farm to School programs in Oregon bring fresh local agriculture into school lunchrooms. Oregon state signed a law (2008) that the Department of Education will put into practice the Oregon Farm-to-School and School Garden Program to promote school district participation in the USDA National School Lunch Program and School Breakfast Program. The program provides information to school districts on how farm-to-school and school garden projects help implement wellness policies mandated by the United States

Department of Agriculture.

Washington

The Washington State Legislature recently passed legislation around a Farm-to-School Program at the state level that helps establish a Washington Grown Fresh Fruits and Vegetable Program. The legislation also includes support for Farmers Markets and Farm-to-Food Banks project. The project was brought forward by the Washington Environmental Council, a coalition of 21 environmental organizations, that includes farmers, children's rights advocates, social groups to end poverty and hunger, and teachers/educators.

British Columbia

Environmental Youth Alliance (EYA)

http://www.eya.ca/

Vancouver B.C.-based non-profit organization that facilitates grass roots action amongst urban youth and community partners to preserve and learn about the social and ecological urban environment.

Farm Folk/City Folk

http://www.ffcf.bc.ca/

FarmFolk/CityFolk Society is a non-profit society that works with farm & city to cultivate a local, sustainable food system that improves access between small farms, producers and urban communities.

City Farmer

www.cityfarmer.org

Website offers a large collection of records from community projects about urban farmers in Vancouver and around the world. Started in 1994, City Farmer was the first web site on the Internet to publish information about urban farming.

UBC Vancouver Farm

http://www.landfood.ubc.ca/ubcfarm/index.php

The 24-hectare site is a hub for learning and research at the University of British Columbia in Vancouver, Canada. The farm includes a Land, Food, and Community Garden, with a Faculty of Education-led "Landed Learning" intergenerational teaching and learning project that involves weekly visits by urban students.

B.C. Ministry of Education—Environmental Concepts in the Classroom
http://www.bced.gov.bc.ca/environment_ed/

Access to innovative new curriculum: Environmental Learning and Experience: An Interdisciplinary Guide for Teachers (2007).

Canadian Centre for Policy Alternatives—Climate Justice Project
http://www.policyalternatives.ca/projects/climate-justice-project

Describes climate change and social justice in the province of British Columbia.

Evergreen
http://www.evergreen.ca/en/

An extensive Canadian environmental site with PDF links on school garden projects, including:

B.C. Agriculture in the Classroom Foundation
http://www.aitc.ca/bc/

Provides resources and lesson plans for teachers, including downloadable handouts. Students can learn about food production and transport, and its impacts on the environment.

B.C. Ministry of Education—Environmental Concepts in the Classroom
http://www.bced.gov.bc.ca/environment_ed/

Provides resources for Environmental Learning and Experience: An Interdisciplinary Guide for Teachers (2007).

Native Plant Society of British Columbia
http://www.npsbc.org/

Includes information about gardening with native plants, education and events, and has a link to an interactive online atlas of native plants—http://www.eflora.bc.ca/

Environmental Educators Professional Specialists Association
http://bctf.ca/eepsa/

Provincial teacher specialist association dedicated to promoting and advancing the teaching of environmental education throughout British Columbia.

Walking the Talk Network/The United Nations Regional Centre for Expertise (RCE) on Education for Sustainable Development
http://www.walkingthetalk.bc.ca/

The B.C.-based network is a multi-sectoral research and environmental network of stakeholders that includes universities, community groups, private organizations, civic government and provincial ministries.

REFERENCES

Ableman, M. (2005). Raising whole children is like raising good food: Beyond factory farming and factory schooling. In *Ecological literacy: Educating our children for a sustainable world*. M.K. Stone, Z. Barlow, & F. Capra (Eds.). San Francisco, CA: Sierra Club Books.

Agard, J. (2004). Palm tree king. In *Emergency kit: Poems for strange times*. London, UK: Faber & Faber.

Anderson, M.K. (2005). *Tending the wild: Native American knowledge and the management of California's natural resources*. Berkeley, CA: University of California Press.

Armstrong, J. 2005. En'owkin. In *Ecological Literacy*. M.K. Stone, Z. Barlow, & F. Capra (Eds.). San Francisco, CA: Sierra Club Books.

Babcock, E.B. (1911). Cooperation between the schools and the college of agriculture. Berkeley, CA: The University of California Press. Accessed June 2010 at http://www.archive.org/stream/universityofcali13berkuoft/universityofcali13berkuoft_djvu.txt

Barbara Boxer Webpage (2010). Accessed June 2010 at http://boxer.senate.gov/news/releases/record.cfm?id=221904

Barlow, Z. & Stone, M.K. (2005). Introduction. In *Ecological Literacy*. M.K. Stone, Z. Barlow, & F. Capra (Eds.). San Francisco, CA: Sierra Club Books.

B.C. Ministry of Education (2007). *Environmental learning and experience: An interdisciplinary guide for teachers*. Available on B.C. Ministry of Education Web site. Accessed June 2010 at http://www.bced.gov.bc.ca/environment_ed/

Bell, A.C., & Dyment, J.E. (2006). *Grounds for action: Promoting physical activity through school ground greening in Canada*. Toronto, ON: Evergreen Foundation. Accessed April 2010 at http://www.evergreen.ca/en/resources/schools/research-policy.sn.

Benedict, H.M. (1913). Will school-gardening survive? *Nature-Study Review*. Vol. 9.

Benton-Short, L. & Rennie Short, J. (2008). *Cities and Nature*. New York: Routledge.

Blackstock, M. (2001). Water: A First Nations' spiritual and ecological perspective.

B.C. *Journal of Ecosystems and Management*. 1: 1–14.

Boggs, G.L. (2009). Detroit: City of Hope. Building a sustainable economy out of the ashes of industry. In *These Times*. Accessed November 2009 at http://www.inthesetimes.com/article/4247/detroit_city_of_hope/

Bowers, C.A. (2006). Revitalizing the commons: Cultural and educational sites of resistance and affirmation. Lanham, MD: Lexington Books.

Caldwell, O.W. (1914). *Nature Study Review*. Vol.10. Accessed November 2009 at: http://www.archive.org/stream/naturestudyrevie10ameruoft/naturestudyrevie10ameruoft_djvu.txt

California Department of Education (2002). Kids' publications, funding are newest ingredients in state's recipe for improved nutrition. California Department of Education Newsletter. Accessed May 2008 at http://www.cde.ca.gov/nr/ne/yr02/yr02rel35.asp

Canadian Council on Learning. (2010). Meeting the demand for trained personnel in Canada's environmental sector. *Learning Link*. Canadian Council on Learning Website. Accessed March 2010 at http://www.ccl-cca.ca/CCL/Reports/LessonsInLearning/LinL2010GreenCareers.htm

Capra, F. (2005). Preface: How nature sustains the web of Life. In *Ecological literacy: Educating our children for a sustainable world*. M.K. Stone, Z. Barlow, & F. Capra (Eds.). San Francisco, CA: Sierra Club Books.

Carr, M. (2008). Geography, spirituality and social change in Cascadia: A bioregional perspective. In *Cascadia the elusive Utopia: Exploring the spirit of the Pacific Northwest*. D. Todd (Ed). Vancouver, B.C.: Ronsdale Press.

Center for Ecological Literacy (2010). Website accessed June 2010 at http://www.ecoliteracy.org/discover/competencies

Churchill, J.A. (1914). From the course of study for the elementary schools of Oregon. *Nature Study Review*. 10:276. Accessed November 2009 at: http://www.archive.org/stream/naturestudyrevie10ameruoft/naturestudyrevie10ameruoft_djvu.txt

City of Seattle. (2009). P-Patch community gardens. Department of Neighborhoods Website accessed November 2009 at http://www.seattle.gov/Neighborhoods/ppatch/)

Cobb, E. [1977]. (2004). *The ecology of imagination in childhood*. Dallas, TX: Springer.

Conan, M. and Quilter, J. (Eds). (2008). *Gardens and cultural Change: A pan-American perspective*. Dumbarton Oaks.

Concise History of Visitacion Valley. *Visitacion Valley Grapevine website*. Accessed November 2009 at http://www.visvalleygrapevine.com/vvvalleyhist.html

Connelly, P. and Ross, C. (2007). Farming the concrete Jungle. In *These Times* website accessed November 2009 at: http://www.inthesetimes.com/article/3297/farming_the_concrete_jungle/Features\

Cox, C., Epps, E., Fishtrom, J., et al. (2005). *Visitacion valley history project*. San Francisco, CA: Arcadia Publishing.

Cranz, G. (1982). *The politics of park design*. Boston, MA: MIT Press.

Davis, B.M. (1905). School gardens for California schools: A manual for teachers. Sacramento, CA: Superintendent State Printing.

Davis, B.M. (1901). Nature study in the Los Angeles state normal. In The Los Angeles normal exponent.

Delgado, M. & Baron, E. (1998). Murals in Latino communities: Social indicators of community strengths *Social Work*. Jul. 1998.

Desmond, D., Grieshop, J., & Subramaniam, A. (2004). *Revisiting garden based learning in basic education*. Food and Agricultural Organization (FAO) of the United Nations, Rome. Accessed November 2009 at www.fao.org/sd/erp/revisiting.pdf

Donald, D. (2004). Edmonton pentimento: Re-reading history in the case of the papaschase Cree. *Journal of the Canadian Association for Curriculum Studies*. 2:1

Dyment, J.E. (2005). *Gaining ground: The power and potential of school ground greening in the Toronto district school board*. Toronto, ON: Evergreen Foundation. Accessed April 2010 at http://www.evergreen.ca/en/resources/schools/research-policy.sn.

Evergreen Foundation Report (2000a). *Stewards and storytellers: The greening of British Columbia school grounds*. Accessed June 2010 at http://www.evergreen.ca/en/lg/stewards.pdf

Evergreen Foundation Report (2000b). *Nature nurtures: Investigating the potential of school grounds*. Vancouver, B.C.: Evergreen. Accessed November 2009 at http://www.evergreen.ca/en/lg/nurtures-en.pdf

Evergreen Foundation Report (2006). *The outdoor classroom. The learning grounds newsletter on school grounds transformation*. 18:12. Accessed November 2009 at http://www.evergreen.ca/en/lg/lg-resources.html

Flanagan, C. (2010). Cultivating failure: How school gardens are cheating our most vulnerable students. *Atlantic Magazine*. January-February. Accessed February 2010 at http://www.theatlantic.com/magazine/archive/2010/01/cultivating-failure/7819/

Foner, P.S. (1979). Introduction. In *On education: Articles on educational theory and pedagogy, and writings for children from the age of gold*. New York:Monthly Review Press.

Francis, M., Cashdan, L. & Paxson, L. (1984). *Community Open Spaces*. Washington, DC: Island Press.

Francis, M. & Hester, H.T. Jr. (1990). *The Meaning of Gardens*. Cambridge, MA: MIT Press.

Freire, P. 1998. *Pedagogy of freedom: Ethics, democracy and civic courage*. Lanhman, MD: Rowman and Littlefield.

Gaylie, V. (2009). The learning garden: Ecology, teaching and transformation. New York: Peter Lang Publishing.

Giesbrecht, S. 2008. The 100-mile curriculum: Place as an educative construct. *Education Canada*. 48:2.

Gottlieb, R. (2005). Epilogue. In *Land of sunshine: An environmental history of metropolitan Los Angeles*. W. Deverell and G. Hise (Eds). Pittsburgh, PA: University of Pittsburgh Press.

Graham, H., Lane Beall, D., Lussier, M. Mclaughlin, P., & Zidenberg-Cherr, S. (2005). Use of school gardens in academic instruction. *Journal of Nutrition Education and Behavior*. 37:3.

Grant, T., & Littlejohn, G. (2004). (Eds). *Teaching green, the middle years: Hands-on learning in grades* 6–8. Gabriola Island, BC: New Society Publishers.

Greene, M.L. (1910). *Among school gardens*. New York, NY: Charities Publication Committee.

Gross, J. (1992). A jail garden's harvest: Hope and redemption. *New York Times*. September 3, 1992. Accessed November 2009 at http://www.nytimes.com/1992/09/03/us/a-jail-garden-s-harvest-hope-and-redemption.html?pagewanted=all

Hale, A. (1998). From "Butterfly." In *Gatherings*. Volume IX: The En'owkin Journal of First North American Peoples. G. Young-Ing (Ed.) Penticton: Theytus Press.

Hodge, C.F. (1914). The Nature Study Course. *Nature Study Review*. 10:276.

Homage to Four Virtues of César Chávez. *Ecology Center*. Accessed November 2009 at http://www.ecologycenter.org/chavez/

Hou, J., Johnson, J.M. & Lawson, L.J. (2009). *Greening Cities, Growing Communities: Lessons from Seattle's Urban Community Gardens*. Seattle, WA: University of Washington Press.

Houghton, E. (2005). *A breath of fresh air: Celebrating nature and school gardens*. Toronto: Sumach Press.

Hucka, J. (2009). Picardo, passion and people: 30 years of p-patching. *City of Seattle Website*.

Jardine, D.W., Clifford, P. & Friesen, S. (2008). *Back to the Basics of Teaching and Learning: Thinking the World Together*. (2nd ed) New York: Routledge.

Jarvis, C.D. (1916). *Gardening in elementary city schools*. Washington, D.C.: Govt. Printing Office.

Jewell, J.R. (1908). *Agricultural education including nature study and school gardens*. Washington, DC: U.S. Bureau of Education.

Jiler, J. (2006). *Doing time in the garden*. Oakland, CA: New Village Press.

Johnson, D. (2005). Pollution and public policy. In *Land of sunshine: An environmental history of metropolitan Los Angeles*. W. Deverell and G. Hise (Eds). Pittsburgh, PA: University of Pittsburgh Press.

Jones, V. (2008). *The green economy*. New York: Harper

Joyce, A. (1914). School gardening in Portland, Oregon. *Nature Study Review*. Vol.10. Accessed November 2009 at: http://www.archive.org/stream/naturestudyrevie10ameruoft/naturestudyrevie10ameruoft_djvu.txt

Kanter, R. (2010). Malik Yakini to speak at Food Access Panel. *DC Food for All*. Accessed April 2010 at: http://dcfoodforall.com/2010/04/malik-yakini-to-speak-at-food-access-panel-tomorrow/

Klein-Miller, L. (1904). *Children's gardens for school and home: A Manual of cooperative gardening*. New York: D. Appleton and Company.

Klemmer, C.D., T.M. Waliczek, and J.M. Zajicek. (2005). Growing minds: The effect of a school gardening program on the science achievement of elementary students. *HortTechnology*, 15 (3) 448–452. Accessed June 2010 at http://horttech.ashspublications.org

Kunstler, J.H. (1993). *The geography of nowhere: the rise and decline of America's man-made landscape*. New York : Simon & Schuster.

Lawson, L. (2005). *City bountiful: A century of community gardening in America*. Berkeley, CA: The University of California Press.

Learning Garden Labs website accessed November 2009 at http://www.pdx.edu/epfa/learning-gardens.html

Levenston, M. (2008). 22 Years later, Lord Roberts school garden. *City Farmer*. Accessed November 2009 at http://www.cityfarmer.info/22-years-later-lord-roberts-school-garden-vancouver-bc/

Lieberman, G.A. and Hoody, L. (1998). *Closing the achievement gap: Using the environment as an integrating context for learning*. Sacramento, CA: CA State Education and Environment Roundtable, 1998. Available at www.seer.org/pages/research

Louv, R. (2008). *Last child in the woods: Saving our children from nature-deficit disorder*. Chapel Hill, NC: Algonquin Books.

The Manufacturer and Builder. (1894). Public School Gardens. 26:9. Accessed November 2009 at http://digital.library.cornell.edu/cgi/t/text/text-idx?c=manu;idno=manu0026–9

Marti, J. [1878]. (1979). *On education: Articles on educational theory and pedagogy, and writings for children from the age of gold*. New York: Monthly Review Press.

Martusewicz, R.A. (2006). Revitalizing the commons of the African-American communities in Detroit. In *Revitalizing the commons: Cultural and educational sites of resistance and affirmation*. C.A. Bowers (Ed.). Lanham, MD: Lexington Books.

Martusewicz, R.A. (2008). Educating for "collaborative intelligence:" Revitalizing the cultural and ecological commons in Detroit. In *Fields of green: Re-storying education*. M. McKenzie, H. Bai, P. Hart & B. Jickling (Eds). Cresskill, NJ: Hampton Press.

Martusewicz, R.A., Lupinacci, J., & Schnakenberg, G. (In Press). *Cultural Studies and environmentalism: The confluence of ecojustice, indigenous knowledge systems and a sense of place*. D. Tippins and M. van Eijck (Eds). New York: Springer Forum Series.

Mendelson, K. (2009). Famous people you never heard of: Notable gardeners, horticulturists, botanists, and landscape gardeners from the Pacific Northwest. *Pacific Northwest Garden History Website* accessed June 2010 at http://www.halcyon.com/tmend/notables.htm

Moore, R.C. and Wong, H.H. (1997). Natural Learning: The life history of an environmental schoolyard. Berkeley, CA: MIG Communications.

Murphy, M. (2003). Education for sustainability: Findings from the evaluation study of the Edible Schoolyard. Berkeley, CA: Center of Ecoliteracy. Accessed November 2009 at http://www.ecoliteracy.org/publications/pdf/ESYFindings-DrMurphy.pdf

Musick, M. (2009). A brief history of Tilth. *Seattle Tilth* website. Accessed November 2009 at http://www.seattletilth.org/about/abriefhistoryoftilth

Orr, D. (2006). The End of education. Lecture at The University of British Columbia. January 13, 2006. Accessed June 2010 at http://www.terry.ubc.ca/index.php/2006/01/27/the-end-of-educationdr-david-orr-jan-12th-13th-2006/

Orr, D.W. (2004). *Earth in mind: On education, environment and the human prospect*. Washington, DC: Island Press.

Orr, D.W. (2005). Place and pedagogy. In *Ecological literacy: Educating our children for a sustainable world*. M.K. Stone, Z. Barlow, & F. Capra (Eds.). San Francisco, CA: Sierra Club Books.

Palmer, C. (1910). Elementary horticulture for California schools. Los Angles state normal school. Accessed November 2009 at: http://www.archive.org/stream/elementaryhorti00palmgoog/elementaryhorti00palmgoog_djvu.txt

Palmer, C. (1910). Elementary Horticulture for California Schools. Los Angles State Normal School. Accessed November 2009 at: http://www.archive.org/stream/elementaryhorti00palmgoog/elementaryhorti00palmgoog_djvu.txt

Palmer, J.A. (1998). Environmental education in the 21st century: Theory, practice and promise. New York: Routledge.

Parsons, H.G. (1910). *Childrens gardens: For pleasure, health and education*. New York: Sturgis & Walton.

Penner, S. (2000). Grandview U'uqinak'uuh community school yard. *Urban Agriculture Notes*. City Farmer Web Site accessed June 2010 at http:// www.cityfarmer.org/grandview.html#grandview

Pennington, G. (1988). Education and school gardens. *City Farmer*. Accessed July 2010 at http://www.cityfarmer.org/scheductn18.html

Pevec, I. (2002). *Ethnobotany: Patterns in relationships*. Evergreen Foundation: Toronto. Available at: http://www.evergreen.ca/en/lg/lg-resources.html

Pevec, I. (2000). *Spirit of nature: Transforming an inner city school grounds*. Unpublished Masters Thesis.

Price, J. (2005). Thirteen ways of seeing nature in Los Angeles. In *Land of sunshine: An environmental history of metropolitan Los Angeles*. W. Deverell and G. Hise (Eds). Pittsburgh, PA: University of Pittsburgh Press.

Purdy, L.E. (1902). Lake Merced wild garden. *San Francisco Chronicle* April 13. Accessed November 2009 at http://foundsf.org/index.php?title=Lake_Merced_Wild_Garden

Riedelsheimer, T. (2002). *Rivers and Tides*. Germany: Mediopolis Film- und Fernsehproduktion.

Rosenblatt, S. and Lin, S. (2005). After the rains, tiny rainbows. *Los Angeles Times*. April 1, 2005. Accessed November 2009 at http://articles.latimes.com/2005/apr/01/local/me-butterflies1

Rowe, S. (2006). *Earth alive: Essays on ecology*. Edmonton, Canada: NeWest Press.

Royal Horticultural Society (2010). Gardening in Schools: A vital tool for children's learning. *National Foundation for Education Research report*. Accessed June 2010 at: http://apps.rhs.org.uk/schoolgardening/teachershome/news/researchonschoolgardening.aspa

Sackman, D. (2005). A garden of worldly delights. In *Land of sunshine: An environmental history of metropolitan Los Angeles*. W. Deverell and G. Hise (Eds). Pittsburgh, PA: University of Pittsburgh Press.

San Francisco Neighbourhood Parks Council (2010). Websited accessed June 2010 at: http://www.sfnpc.org/visvalleygreenway

Schiffman, P. (2005). The Los Angeles prairie. In *Land of sunshine: An environmental history of metropolitan Los Angeles*. W. Deverell and G. Hise (Eds). Pittsburgh, PA: University of Pittsburgh Press.

Shibley, M.A. (2008). The promise and limits of secular spirituality. In *Cascadia the elusive Utopia: Exploring the spirit of the Pacific Northwest*. D. Todd (Ed). Vancouver, B.C.: Ronsdale Press.

Shiva, V. (2005). *Earth democracy: Justice, sustainability and peace*. Cambridge, MA: South End Press.

Smit, J. (2006). Urban agriculture: Improving the environment for living and contributing to a sustainable ecosphere. *City Farmer* Website. http://www.cityfarmer.org/deskSmit.html

Smith, A. & MacKinnon, J.B. (2007). *The 100-mile diet*. Toronto, ON: Random House Canada.

Smith, G. (1999). Creating a public of environmentalists: The role of non-formal education. In *Ecological education in action: On weaving education, culture and the environment*. G. A. Smith and D. R. Williams (Eds). New York: SUNY Press.

Smith, L.L. & Motsenbocke, C. (2005). Impact of hands-on science through school gardening in Louisiana Public Elementary Schools. *HortTechnology*, 15 (3) 439–443. Accessed July 2010 at http://horttech.ashspublications.org

Sneed, C. (2000). Is it time to close the prisons? *Yes Magazine*. Fall 2000. Accessed November 2009 at http://www.yesmagazine.org/article.asp?ID=381

Snyder, G. (1990). *The practice of the wild*. San Francisco, CA: North Point Press.

Spirn, A.W. (1984). *The granite garden: Urban nature and human design*. New York: Basic Books.

Starr, K. (2006). *Coast of dreams: California on the edge, 1990–2003*. New York: Vintage Books.

Stebbixs, C.A. (1912). Growing children in california gardens. *Nature Study Review*. 8: 2 http://www.archive.org/stream/naturestudyrevie08ameruoft/naturestudyrevie08ameruoft_djvu.txt

Stine, S. (1997). *Landscapes for learning: Creating outdoor environments for children and youth*. New York: John Wiley & Sons.

Stone, M.K. (2005). Sustainability: A new item on the lunch menu. In *Ecological literacy: Educating our children for a sustainable world*. M.K. Stone, Z. Barlow, & F. Capra (Eds.). San Francisco, CA: Sierra Club Books.

Subramaniam, A. (2002). A garden based learning in basic education: A history. *University of California monograph*. 4H Centre for Youth Development: University of California, Davis.

Tezozomanc (2006). Sustainability and Alternatives. *Free Speech TV*. Season 3:Episode 9. Accessed February 2010 at http://www.archive.org/details/freespeechtv_sourcecode3_08

Todd, D. (2008). The promise and limits of secular spirituality. In *Cascadia the elusive utopia: Exploring the spirit of the Pacific Northwest*. D. Todd. (Ed). Vancouver, B.C.: Ronsdale Press.

Trelstad, B. (1997). Little machines in their gardens: A history of school gardens in America, 1891 to 1920. *Landscape Journal*. 16:2.

Trinkl, J. Jack London: Historical Essay. On *Found SF*. Accessed November 2009 at http://foundsf.org/index.php?title=JACK_LONDON

Warner, S. B. (1987). *To dwell is to garden*. Boston, MA: Boston University Press.

Washington, B.T. (1911). My larger education. *Chapters from my experience*. http://docsouth.unc.edu/fpn/washeducation/washing.html

Waters, A. (2008). *The edible schoolyard: A universal idea*. Berkeley, CA: Chronicle Books.

Weed, C.M. (1909). *The school garden book*. New York: Charles Scribner's & Sons.

Welby, S. (2003). Patterns through the seasons: A year of school food garden activities. Victoria, British Columbia: *Evergreen/Lifecycles*. Accessed November 2008 at: www.evergreen.ca/en/lg/patterns.pdf

Wexler, M. (2008). Conjectures in workplace Spirituality in Cascadia. In *Cascadia the elusive utopia: Exploring the spirit of the Pacific Northwest*. D. Todd. (Ed). Vancouver, B.C.: Ronsdale Press.

Williamson, E. (2002). A deeper ecology: Community gardens in the urban environment. Analytical paper submitted to urban affairs and public policy of the University of Delaware. Retrieved November 2009 at http://www.cityfarmer.org/AP—city_farmer.doc

Williams, D.R., Taylor, S. (1999). From margin to centre: Initiation and development of an environmental school from the ground up. In *Ecological education in action: On weaving education, culture and the environment*. G. A. Smith and D. R. Williams (Eds). New York: SUNY Press.